Coastal zone planning and management

Coastal zone planning and management

Proceedings of the conference *Coastal management '92: integrating coastal zone planning and management in the next century,* organized by the Institution of Civil Engineers and held in Blackpool on 11-13 May 1992

Thomas Telford, London

Conference organized by the Institution of Civil Engineers and co-sponsored by the Ministry of Agriculture, Fisheries and Food, the Royal Institution of Chartered Surveyors and the National Rivers Authority

Organizing Committee: M. G. Barrett, Posford Duvivier (Chairman); I. R. Whittle, formerly National Rivers Authority; Dr S. W. Huntington, HR Wallingford; Dr C. A. Fleming, Sir William Halcrow & Partners; L. Leeson, Countryside Commission; R. G. Purnell, Ministry of Agriculture, Fisheries and Food; T. M. Cox, Sefton Metropolitan Borough Council

A CIP catalogue record for this publication is available from the British Library

ISBN 0 7277 1904 1

First published 1992

The Authors and the Institution of Civil Engineers, 1992, unless otherwise stated

All rights, including translation, reserved. Except for fair copying, no part of this publication may be reproduced, stored in a retrieval system or transmitted in any form or by any means electronic, mechanical, photocopying, recording or otherwise, without the prior written permission of the Publications Manager, Publications Division, Thomas Telford Services Ltd, 1 Heron Quay, London E14 4JD.

Papers or other contributions and the statements made or opinions expressed therein are published on the understanding that the author of the contribution is solely responsible for the opinions expressed in it and that its publication does not necessarily imply that such statements and or opinions are or reflect the views or opinions of the organizers or publishers.

Published on behalf of the organizers by Thomas Telford Services Ltd, Thomas Telford House, 1 Heron Quay, London E14 4JD.

Printed in Great Britain by Redwood Press Ltd, Melksham, Wilts.

Contents

Opening address. LORD HOWE	1

Setting the scene
1.	The development of coastal engineering. C. A. FLEMING	5
2.	Coastal conservation. R. W. G. CARTER	21
3.	Setting the scene. G. A. D. KING	37
Discussion		47

Pressures on the zone
4.	Natural change. J. PETHICK	49
5.	Coastal heritage. M. KIRBY	65
6.	Regenerating the developed coast. L. SHOSTAK	71
Discussion		81

Coastal heritage
7.	Marine nature conservation in the coastal zone. S. GUBBAY	83
8.	The open coastline. C. STEVENS	91
9.	Estuaries. P. I. ROTHWELL	101
Discussion		113

Development pressures
10.	Development pressures in an area of rising human resources. M. N. T. COTTELL	115
11.	Development pressures in an area of declining human resources. M. J. HORSLEY	125
12.	Cardiff Bay - microcosm of conflicts. D. A. CROMPTON	139
Discussion		143

Framework for planning and management
13. Coastal planning: recent policy developments. J. ZETTER — 145
14. Coastal defence: legislation and current policy. J. R. PARK — 153
15. Overseas responses through legislation and policy to coastal problems. J. C. DOORNKAMP — 161
Discussion — 179

Planning options in response to change
16. Coastal zone planning. T. M. COX — 181
17. Shoreline management: a question of definition. I. H. TOWNEND — 195
18. Coastal zone planning beyond the year 2000. I. R. WHITTLE — 211
Discussion — 219

Engineering sympathetic solutions
19. Beaches - the natural way to coastal defence. A. H. BRAMPTON — 221
20. Engineering the beaches. G. M. WEST — 231
21. Engineering with conservation issues in mind. K. A. POWELL — 237
Discussion — 251

Case studies
22. The Anglian Management Study. M. W. CHILD — 253
23. The defences for Lincolnshire. R. S. THOMAS — 269
24. The provision of marine sediments for beach recharge. A. J. MURRAY — 283
Discussion — 291

Conclusions
25. Difficult decisions. K. J. RIDDELL — 293
26. Coastal groups. T. A. OAKES — 311
Discussion — 323

Closing address. M. G. BARRETT — 325

Opening address

LORD HOWE, Parliamentary Secretary for Agriculture, Fisheries and Food

As an island race, the British cannot afford to lose sight of how vitally important the state of their coastline is - for a whole host of reasons - economic, environmental and recreational. No-one in the UK lives more than 84 miles from the coast, and increasingly people want to live as close to it as possible. Coastal areas are subject to a range of conflicting pressures, including the most inexorable pressure of all - the sea itself. This Conference addresses many of these conflicting pressures and interests. These topics are also examined in a report on coastal zone protection and planning by the House of Commons Environment Select Committee, which the Government is studying carefully and to which MAFF will give formal response in due course. The discussions at this Conference are timely and important.

My Department has a number of key responsibilities related to the coastline including, of course, fisheries management, regulatory powers over the disposal of wastes at sea and, most relevant to this Conference, flood defence and coast protection.

The damage the sea can do must be kept in mind. In the 1953 floods, which were caused by the combination of a surge tide and gale force winds, waves smashed through sea walls and rivers overtopped their banks. Many of East Anglia's coastal areas were devastated. The human cost was tragic: 300 lives were lost and many more people were affected by what they had suffered - 24 000 houses were damaged, 200 industrial properties were put out of action, 11 trunk roads were rendered impassable, 160 000 acres of agricultural land were left sterile for at least a year, 13 000 livestock and 34 000 poultry were killed. This is what the sea can do, and this is why I believe that it is important to co-ordinate development along coasts carefully with the measures that can be taken to defend them.

This Government has shown a continuing commitment to the defence of coasts against the ravages of the elements by substantially increasing planned grant expenditure. Over six years there has been a 100% increase in grant provision for flood defence and coast protection - from £31 million in 1986-87 to £62 million for 1992. This is the largest budget ever.

Given the scale of the problem and the resources required to deal with it, I think it must be agreed that issues of flood defence and coast protection should be tackled strategically rather than piecemeal. That sort of approach provides a better opportunity to gauge the larger issues such as the effects of geographical features, the action of the tides on coasts and the results of global warming, and to assess the environmental consequences of alternative courses of action locally and nationally.

MAFF has overall policy responsibility for flood defence and coast protection and is empowered to pay grant to the National Rivers Authority, to local authorities, and to Internal Drainage Boards in respect of works that meet three standards: they have to be technically sound, economically worthwhile and environmentally sympathetic. This is at the operational level. In December 1991 the Minister of Agriculture, John Gummer, announced that work was under way on the development of a national strategy for flood and coastal defences. MAFF intends to consult widely with flood defence agencies and conservation bodies over the scope of that strategy. It will be a strategy based on the best scientific understanding available, and will have environmental considerations as a central point of reference.

This strategic approach to coastal defence is the reason why MAFF has been encouraging local authorities to consult more closely with each other and to establish closer links with the NRA and the Association of District Councils. This has led to the setting up of Coastal Groups, which now cover about 98% of the coast of England and Wales. The members of these Groups include not only technical experts, such as engineers from the authorities concerned, but also other interests such as representatives of conservation bodies and private land-owners. This is a good, practical way of co-ordinating action and sharing information. The individuals in the Groups bring a wealth of expertise and local knowledge to bear on the problems affecting the coastline - and are yet another way in which long-term planning can be fostered. Since 1991 these Groups have regularly met in a national Coastal Defence Forum, chaired by MAFF, which allows all involved to share their concerns and their expertise.

On the subject of the disasters in East Anglia MAFF provided nearly £1 million in grant towards the costs of the Anglian Sea Defence Management Study initiated by the NRA. This will not only be of practical benefit to local authorities in that region as they plan and design coastal defence work, but will also foster sensible long-term solutions and encourage all interested parties to develop more environmentally acceptable and appropriate responses to sea defence problems. It will be vitally important for the study's database to be kept up to date and for it to be made available to everybody. In time I hope it will be possible for other stretches of the coastline to be examined in a similar way.

Increasingly more emphasis is being given to the environmental aspects of flood and coastal defence works. With care and understanding, such works need not have an adverse impact on flora and fauna, on historic and archaeological remains, or on the landscape; in some cases they can even be enhanced. Wetlands, rivers, marshes, coasts and estuaries are especially vulnerable and there could also be effects on the amenity value of an area. New and improvement works must comply with environmental assessment procedures. In addition, in order to avoid any detriment to the environment, arrangements exist to ensure that the full range of environmental interests are drawn into the early planning stages of schemes, so that decisions about design and construction take account of those interests. The *Conservation Guidelines for Drainage Authorities*, issued by MAFF jointly with the Department of the Environment and the Welsh Office, provide useful guidance on the statutory requirements and emphasise the need for authorities to consult environmental bodies about proposed works. These guidelines have recently been updated and strengthened in consultation with conservation bodies and operating authorities.

Furthermore, as a result of another initiative of John Gummer, MAFF is commissioning the preparation of two advisory publications for use by coastal defence authorities. These will provide detailed guidance as to how operating authorities can carry forward their programmes with even greater sympathy for environmental concerns - the protection of birds and wildlife and the need to look at the more natural ways of defending the coastline, notably by the replenishment of beaches with new shingle.

The Government's commitment to the environmental cause is also evidenced by MAFF's financial input in other directions. MAFF commissions relevant research and disseminates the results widely. It grant aids eligible costs of preparing environmental statements arising from compliance with the environmental assessment procedures and also grant aids the cost of works necessary to make schemes sympathetic to the environment.

In recent years all concerned with flood defence and coastal protection have become a lot better at providing the protection needed for essential urban, industrial and agricultural areas in ways which are sympathetic to the environment, and as understanding of natural processes develops, no doubt it will be possible to do even more. All the time there are new challenges, such as global warming and rising sea levels, for policy. Sound science, more reliable predictions and more extensive understanding of natural processes will continue to underpin coastal defence policy. Currently the Ministry spends about £2.6 million a year on research in this field, of which over £1 million a year is devoted to coastal processes and alternative means of protection. This is by any standard a significant commitment of resources.

1. The development of coastal engineering

C. A. FLEMING, PhD, FICE, Director, Sir William Halcrow & Partners Ltd

Coastal works to protect people and property have been built by engineers for hundreds of years in the UK, yet it is probably only in the last half-century that a real appreciation of coastal processes has emerged. The present legislative framework has also developed over a long period and is both complex and involved with a great many interested parties. More recent developments have seen the establishment of formal shoreline management procedures together with the voluntary formation of Coastal Groups covering most of the coastline in England and Wales. Although in its infancy this provides a sound base from which to evolve a Coastal Management framework which addresses rather wider issues concerning planning, sustaining resources and enhancing the environment within the coastal zone.

Introduction

The coastline has been 'engineered' for many centuries, initially for the development of ports and maritime trade or fishing harbours to support local communities. Early sea defences were embankments, but when dealing with coastal erosion problems the hard edge approach dominated at least in the United Kingdom. In particular the Victorians were active in their desire to construct promenades in seaside resorts which were usually vertically faced. Coastal processes were not only poorly understood, but there was some confusion as to what the driving forces are. It was not until the post Second World War period that the theoretical models and ideas that underlie the basic processes began to be developed, save for basic wave and tidal motion.

The value of attempting to retain beach material, whether for sea defence, coast protection or recreational use has been recognised for some time. This is to some extent demonstrated by the extensive lengths of coastline that have been groyned in the past. However, it has been suggested that the responsible Authorities have, quite naturally, dealt with these matters on a parochial basis with little regard for, or appreciation of, the impact of their actions on neighbouring territory.

This has allegedly led to some rather undesirable consequences in both conservation and planning terms and the Engineer has been criticised for being insensitive and not paying heed to these issues. There are a number of other factors that should be taken in account before coming to this conclusion. These include the constraints that have, in effect, been imposed by interpretation of Government legislation and the nature of the responsibilities that fall upon the various Authorities involved in implementing coastal works. These have primarily been to protect people and property from the effects of erosion or flooding in situations where economic justification can be established. In this regard they have generally been demonstrably successful.

It is also evident that the planning system has not generally taken the question of long term coastal evolution into account, when in many instances planning permission has been granted for development on sites that have been well known to be vulnerable to long term erosion. At the same time conservation issues have developed alongside our appreciation of natural processes and the complex interactions involved.

The science that underpins near shore coastal processes and hence engineering appreciation is relatively young in its development, having only emerged as a subject in its own right over the past fifty years. During that time there have been rapid advances in knowledge and understanding, thus allowing solutions to coastal problems to become very much more sophisticated with respect to harmonisation with the natural environment. There has thus been an evolution of design practice that has progressively been moving towards 'softer' engineering solutions. That is, those solutions which attempt to have a beneficial influence on coastal processes and in doing so improve the level of service provided by a sea defence or coast protection structure.

The most recent developments have led to the adoption of more formal shoreline management practice. A careful distinction is made here to avoid confusion with coastal (zone) management. A notable lead in the UK was taken by the Anglian NRA (formerly The Anglian Water Authority) when they embarked in 1987 on a comprehensive study over the entire length of the coastline for which they are responsible with the overall objective of developing a shoreline management strategy to meet both short term and long term needs.

There has also been the formation of some 18 coastal cell groups which act as voluntary co-ordinating bodies over much of the coastline of England and Wales. In some instances these are funding research into coastal processes for the mutual benefit of all of the members of the group and thus foster a reciprocal appreciation of the problems involved.

Historical perspectives

It is said that the design and construction of man-made harbours is one of the oldest branches of engineering. In his Presidential Address in 1940

Savile (ref. 1) mentions the Port of A-ur built on the Nile prior to 3000 BC and nearby on the open coast the Port of Pharos around 2000 BC. The latter had a massive breakwater more than 2.5 km long. The Romans invented a hydraulic cement and developed the practice of pile driving for cofferdam foundations, a technique that was used for the construction of concrete sea walls. Whilst these structures were no doubt built on the basis of trial and error procedures there is no evidence that there was any real appreciation of coastal processes with respect to siting of maritime infrastructure.

Philpott (ref. 2) reviewed a number of historic texts in order to establish when the question of whether alongshore transport is caused by waves, or by currents, or by some combination of the two, was really established. It makes interesting reading

> "For at least a century prior to 1930, this was subject of a great debate in British and US engineering societies. Sides were taken. Harbour engineers felt obliged to state their position. For example, in five major papers and the ensuing discussions that were reported between 1885 and 1914, forty-seven engineers stated their position, most of them more than once. Of those, fifteen held that waves alone caused alongshore sediment transport, twenty-five that it was currents alone, and seven both. Textbook writers tended to equivocate. The five papers were: Vernon-Harcourt (1882) (ref. 3), Haupt (1899) (ref. 4), Carey (1904 and 1905) (refs 5 and 6) and Spring (1914) (ref. 7). This controversy did not merely surround the subject of coastal processes, it became the core of the subject. It was the competition between two paradigms. The subject could not make significant progress until a clear winner was recognised.
>
> Although Sir Francis Spring with the last mentioned paper, his classic study 'Coastal Sand Travel at Madras Harbour', should have settled the debate, the confusion lingered on. In 1928, Brysson Cunningham, in the third edition of his standard and widely accepted text on harbour engineering was still caught up somewhere in the middle of the dilemma. Though he tended to favour waves, and he quotes from Spring, he did not fully understand the implications of the new beach coastal process model. That was still not the end of it, the shadow of the debate, much weakened, can be traced as far forward as Cornick (1969) (ref. 8), some fifteen years after the publication of TR4. Then it was forgotten.
>
> The historical process described above may be seen as an unusual example of a 'scientific revolution' in an infant science, Kuhn (ref. 9); unusual in the sense that it was both prolonged and apparently unnecessary. It was unnecessary because earlier eminent harbour engineers like Palmer (1834) (ref. 10) and Rennie (1854) (ref. 11) very clearly seemed to understand the primary role of wave action..."

SETTING THE SCENE

There have been several periods of development of coastal works in the UK over the past half-century. There was an extensive wall building programme during the 1930's as part of the unemployment relief schemes. These were based on dock wall designs with near vertical profiles. The development of the Mulberry Harbours in the Second World War led to the concept of determining wave climate, using wind data and design parameters such as wave height and wave period.

Thus contemporary coastal engineering effectively began at that time witnessed by the First Conference on Coastal Engineering at Berkeley, California sponsored by The Engineering Foundation Council on Wave Research (USA). This was closely followed in 1954 with the publication and widespread acceptance of 'Shore Protection, Planning and Design -Technical Report No. 4' (TR4) by the US Army Corps of Engineers, Beach Erosion Board. One wonders if it was merely an accident that the 'Planning' part of the title was later dropped and it became the well known 'Shore Protection Manual'.

The history books are full of accounts of major storms that caused destruction and devastation to various sections of the coast. In recent times one of the most significant dates in coastal engineering in England is 31 January 1953 when an extreme storm surge travelled down the North Sea coincidentally with extreme storm waves. The effect was devastating and serves as a poignant reminder as to how vulnerable the low lying areas of the East Coast are. The post 1953 period saw great activity in the construction of sea defences along that coastline at a time when sea walls and groyne systems were the norm and the overriding criterion was to provide a secure safety barrier against any such event occurring again.

By the 1960's a much greater understanding of coastal processes emerged as the theoretical development coupled with physical and numerical modelling developed. This led to a gradual re-appraisal of coastal engineering techniques in such a way that design processes began to consider studies of the coastal regime and its interaction with the proposed works. By the early 1970's this led to the application of relatively novel solutions to coastal problems such as beach nourishment, artificial headlands, offshore breakwaters etc.

The legislative framework

Coastal Works in England and Wales are administered under two main Acts of Parliament: the Coast Protection Act 1949 and the Land Drainage Act 1976. The former relates to the protection of land against erosion and the works are known as 'Coast Protection', whilst the latter is applied to land liable to flooding and the works are known as 'Sea Defences'. The responsibility for carrying out such works falls upon local authorities such as District

Councils and the newly formed National Rivers Authority, save for those areas that are either privately owned, administered by the Ministry of Defence etc. However, District Councils do also have powers under the Land Drainage Act 1976 in appropriate circumstances. The Ministry of Agriculture, Fisheries and Food (MAFF) is responsible for administering the Acts in England, whilst the Secretary of State for Wales is responsible in that country. A summary of some of the primary Acts applicable in England and Wales is given in Table 1, modified from Townend (1986) (ref. 12).

Similar arrangements apply to Scotland and Northern Ireland through the Land Drainage (Scotland) Act 1958, the Flood Prevention (Scotland) Act 1961, the Coast Protection Act 1949 and in Northern Ireland the Drainage Order Act 1973.

A comprehensive account of legislation and policy, the history leading to present legislation, the way in which it applied to different parts of the coastline and the requirements underlying the approval of schemes and associated grant aid is given by Park (1989) (ref. 13) in the previous ICE Coastal Management Conference. He describes how for sea defences 49 Catchment Boards were set up in 1930 to be reduced to 34 River Boards in 1948, 27 River Authorities in 1963 and 10 Regional Water Authorities in 1973, latterly partially transformed into the National Rivers Authority in 1989 due to privatisation of the water and sewerage utilities. However, the authorities are still required to carry out their drainage, flood and sea defence responsibilities through Regional Land Drainage Committees, now called Flood Defence Committees.

A common criticism made with respect to coastal works legislation in the UK is that it is too complex with many different Acts relating to a variety of activities with different and competing interests. Regulatory controls are inconsistent and there are a multitude of public authorities that either administer the Acts or have interests within the coastal zone and thus, quite rightly, require consultation. These arrangements do not lend themselves towards a unified and coherent coastal zone strategy. Consequently there are often calls for one National body responsible for planning and policy for all matters within the coastal zone. However, this is not a simple matter.

In 1985 a review of separate coastal protection and sea defence schemes led to a Green Paper on Financing and Administration of Land Drainage, Flood Prevention and Coast Protection by MAFF and proposed integrating arrangements under a redefined category of 'coastal works'. This proposal was not well received by the Maritime District Councils as it would have put the overall responsibility with the Water Authorities with the provision that the MDCs could be designated as the Coastal Works Authority over their particular lengths of coastline. At the same time the County Councils wished to retain strategic control of coastal policy. This indicated just some of the difficulties that arise in attempting to effect an integration policy. However,

SETTING THE SCENE

Table 1. Primary legislation applying to the coastal zone in England and Wales

LEGISLATION	PURPOSE
Coast Protection Act, 1949	Prevention of erosion and encroachment by the sea.
Land Drainage Act, 1976	Navigation, Sea Defences to prevent flooding.
Food and Environment Act, 1985	License to construct coastal works with respect to pollution
Water Act, 1989 superseded by Water Resources Act, 1991	Transfer of responsibility to the National Rivers Authority
Environmental Protection Act, 1990	Integrated Pollution Control
EC Directives (numerous)	Coastal water quality and environmental issues
Crown Lands Act, 1866	Mining, quarrying and dredging.
Wildlife and Countryside Act, 1981	Conservation of marine flora and fauna.
Sea Fisheries Regulation Act, 1966	Control of fishing industry.
Control of Pollution Act, 1974	Restriction of discharge into coastal waters.
Dumping at Sea Act, 1974	Prohibition of dumping in territorial waters.
Prevention of Oil Pollution Act, 1971	Prohibition of discharge of oil in territorial waters
Highways Act, 1959	Quarrying and navigation.

whilst such integration would simplify some matters regarding shoreline management it would not address the much wider issues related to coastal zone management. Townend (1992) (ref. 14), in these conference proceedings, makes some interesting proposals for a management framework within which Shoreline Management at a Regional level becomes a component of Coastal Zone Management at a National level by adjusting rather than overturning the existing administrative arrangements.

Evolution of design practice

The development of understanding of coastal processes has been described above. The consequences of 'bad design' by building a hard edge structure on a shoreline was however appreciated. An article written by T. B. Keay (1941) (ref. 15), reproduced in The Dock and Harbour Authority notes that "The efforts of man to prevent erosion are sometimes the cause of its increase, either at the site of his works or elsewhere along the coast". This he illustrated as shown in Fig. 1 with an example of a sea wall built at Scarborough in 1887. In just 3 years it was necessary to add an apron and in a further 6 years an additional toe structure and timber groynes. He went on to say that an essential preliminary of all coast protection works is to study the local natural conditions. These he described, but again made no mention of waves. There are many other examples that could have been cited.

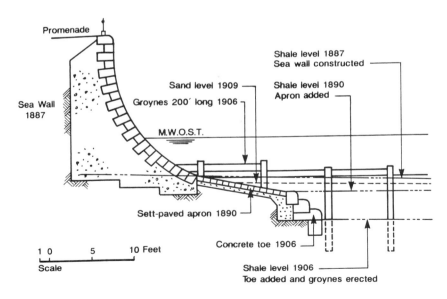

Fig.1. Seawall at Scarborough

SETTING THE SCENE

The major influences that coastal works have had on the shoreline centres on the degree of interference with the natural processes that are taking place. Harbours and their approach channels have had a significant impact on alongshore drift as have coastal defences themselves through the use of groynes or other similar structures. It is also evident that protection of some types of coast from erosion must deprive the local and adjacent beach system of some of its natural sediment supply. Given that nature will always try to re-establish some form of dynamic equilibrium that shortfall in supply is redressed by removing material from elsewhere. That situation can also be exacerbated by introducing structures that, instead of absorbing energy as a natural beach does, reflect the incident waves to do more damage on the beach in front of the wall. Other processes that may be interfered with include the overwash of coastal banks and the natural roll-back of salt marsh systems.

It is indeed well known that successful design requires a thorough investigation of the natural conditions in all spheres of engineering. For coastal works a modern study would typically include

(a) determination of the wave climate (sometimes on the basis of hindcasting from winds), currents, tides, extreme surge levels, transformation of waves as they approach the shoreline and the joint probability of these events

(b) a study of historic information from a variety of sources to determine trends and rates of accretion and erosion

(c) estimates by prediction or measurement of the movement of sediment

(d) on the beach and in the nearshore and establishment of a sediment budget

(e) determination of the influence of coastal works, if any, on the neighbouring sections of coastline and any future trends; similarly the impact that any proposed works might have on both the local and regional regime

(f) evaluation of environmental impact with regard to all of the uses of the shoreline including any impacts on water quality and local ecology as a result of the works

(g) identify all different types of schemes that would provide a possible solution together with possible construction materials, methods and costs

(h) evaluation of the level of service that is required from any capital works for alternative options set against technical advantages, capital and maintenance costs, tangible and intangible benefits and disbenefits both in terms of infrastructure value and environmental issues

(i) evaluation of mitigation measures that might be used to counterbalance any undesirable impacts of the works

(j) consultation with all statutory bodies with interests in the area and the public in general.

Many of the factors described above are highly interactive both in consideration of design development and the definition of the physical parameters that need to be taken into account. The analytical tools and data collection that will be used to help to develop an understanding of a coastal system with respect to the design of works will usually address the topics given in Table 2. Thus historic records, field measurements, numerical models and physical models can all contribute to this process. Clearly if a shoreline has been systematically monitored there are immediate benefits that can be realised. Neither should the value of simple, but informed, observation be undervalued.

There are a wide range of coastal works that might be employed to tackle a particular situation, each of which may perform a number of different functions. They will also have differing engineering life spans as well as different capital and maintenance cost steams. The potential economic benefits can also have a strong influence on the final solution, on the basis of what the project can afford rather than what might be the most desirable with respect to physical processes (ref. 16). Fig. 2 shows some of the more common types of coastal works that are commonly employed today. These include artificial headlands, groynes, offshore breakwaters, beach nourishment and sea walls. Their basic advantages and disadvantages are also listed.

Present-day design practice places much emphasis on attempting to hold a healthy beach on the shoreline as the primary of protection. A sufficiently substantial beach can accommodate the dynamic changes that are the result of differing climatic conditions. These so-called 'soft' solutions are generally considered to more 'environmental friendly' than the traditional 'hard' protection works. However, where human life and conurbations exist one cannot afford to compromise on safety issues and Barrett (ref. 17), points out that a 'hard' element of a protection scheme, which may not necessarily be ordinarily exposed to direct attack, can provide a number of advantages.

Formal shoreline management

The main criticism that has arisen in the past is that, whilst there have been significant advances in the appreciation of the interactions involved in regional coastal processes, the physical areas of responsibility, and hence parochial interest, have constrained studies to sub-areas in the context of offshore and nearshore dynamics. In those circumstances rehabilitation or new works were initiated as and when particular problems arose, sometimes prioritised where significant lengths of coastline were involved. Existing data was generally only available as the result of local ad-hoc monitoring and design and construction would be on the basis of local site specific studies and measurements. The constraints that might be imposed by interactions with adjacent sections of shoreline would only generally be

SETTING THE SCENE

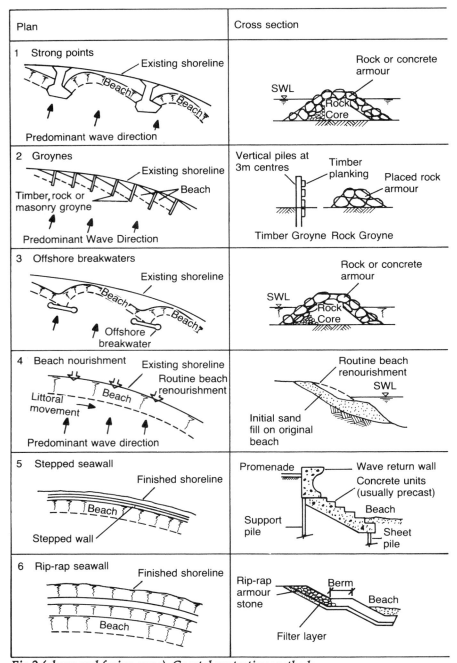

Fig.2 (above and facing page). Coastal protection methods

PAPER 1: FLEMING

Advantages	Disadvantages
1. Acts like a natural headland. 2. Promotes natural beach. 3. Easy to construct. 4. Little maintenance required.	1. Large structures. 2. Can cause erosion downdrift of protected length of coast.
1. Reduces littoral movement. 2. Easy to construct.	1. Regular interruptions in beach. 2. Regular maintenance required. 3. Can induce local scour. 4. Can cause erosion downdrift.
1. Provides stable beach plan forms. 2. Promotes natural beach. 3. Little maintenance required. 4. Can allow controlled littoral movement.	1. Difficult to construct. 2. Large structures. 3. Can cause erosion downdrift.
1. No structures to interrupt beach. 2. Attractive if longshore drift rate is small.	1. Requires regular maintenance. 2. Constant source of sand required. 3. Unlikely to be economic in severe wave climates or where sediment transport is rapid.
1. Wide range of alternative designs. 2. Promenade above as recreation feature. 3. Steps provide easy access.	1. Wave reflection similar to a vertical wall. 2. Scour at toe erodes beach. 3. Regular maintenance required. 4. Does not aid beach stability.
1. Good hydraulic performance. 2. Easy to construct. 3. Little maintenance required.	1. Ramps required for access to beach. 2. Does not aid beach stability.

SETTING THE SCENE

considered in the context of the particular site specific problem, without addressing the wider implications.

This type of framework has been called the 'Traditional Approach' by Townend (ref. 18). He points out that such an ad hoc approach is often unsatisfactory as it makes it extremely difficult to ensure that, not only are schemes developed to be efficient and cost effective for the nation as a whole, but also natural processes and natural resources are used to best effect. The benefits of a more strategic approach to shoreline management should be easily appreciated.

In 1987, the National Rivers Authority, Anglian Region (the Anglian Water Authority) commissioned a shoreline management study to cover one

Table 2. Typical study components and methods of assessment

SUBJECT	METHOD
Deep Water Wave Climate	Field Measurement Numerical Hindcast from Wind Data
Tide Levels and Currents	Field Measurement 2D Numerical Model
Storm Surge	Long Term Water Level Measurements 2D Numerical Model
Shallow Water Wave Climate	Field Measurement Numerical Models - Wave Refraction/Diffraction
Design Criteria	Statistical Analysis of Extremes and Joint Probability Required Level of Service
Sediment Circulation	Historic Analysis Numerical Model - Tidal Residuals Field Measurements - Sediment Path Analysis
Beach Response	2D/3D Physical Model Numerical Beach Plan Shape Numerical Profile Model Numerical 3D Response Model
Run-Up and Overtopping	2D Physical Model Semi-Empirical Calculations
Structural Integrity	2D/3D Physical Model

of the most vulnerable coastlines in Britain stretching from the estuaries of the Humber to the Thames. The general thrust of the investigation was to provide an understanding of the mechanisms causing changes in foreshore levels leading to the development of both short term and long term shoreline management strategies.

A management framework was developed which is reported fully elsewhere (refs 19 and 20) and in general terms in these proceedings (ref. 14). Strategy must be developed by firstly considering the relevant legislation and statutory responsibilities of the Authority. These will form the basis of the Management Objectives. It is then necessary to define Policy Guidelines which indicate which policy options should be applied to a particular situation. These consider both the location of the protection works and the level of service to be provided. The final stage is to define the Management Response Options, which for engineering works will be based on the physical characteristics of the site, the processes taking place there and the nature of what is being protected.

There are a number of engineering and system management options that are possible and include

(*a*) reinstate - where the existing form of defence is appropriate and should be maintained; examples are beach renourishment/recycling, saltings regeneration, structural reconstruction

(*b*) modify - make some adjustment to the existing form of defence either to improve performance or reduce any negative effect; examples are removal of natural features or structures, structural alterations, stabilisation (cliffs/dunes/saltings), nearshore intervention (dredging at inlets and banks/sand bypassing)

(*c*) create - install new structures which can work within the prevailing regime to improve the overall defence system; examples are embayments (offshore breakwaters/headlands/groynes), linear protection (revetments/embankments/seawalls)

(*d*) monitor - where no intervention is planned there will always be a need to monitor through condition and coastal response surveys.

It should be noted that the option to monitor is considered to be the absolute minimum level of shoreline management. It is only by regularising routine monitoring through measurement and observation that real insights into coastal behaviour in response to specific events can readily be realised. It is too late if one waits until something needs to be done before any such measurements are initiated.

The Sea Defence Management Strategy for the Anglian Region is now at the implementation stage so that all new coastal works are reviewed within the framework that has been developed (ref. 21). At the same time a consistent monitoring programme has been set up throughout the region. It is

SETTING THE SCENE

hoped that the District Councils within the region will also participate through initiating their own monitoring programmes and exchanging data with the NRA.

In other parts of the country there have also been significant developments that will assist in co-ordinating activities within the coastal zone. This has come about by the formation of Regional Coastal Groups, which are voluntary bodies who meet on a regular basis to discuss matters of mutual interest. Some groups also fund research projects and other studies that are based on common objectives. In some areas local plans are being developed that do integrate shoreline management into the overall planning framework. It can thus be appreciated that these initiatives resulting from self motivation in the realisation that this type of co-ordination is important, have been rapidly gaining momentum and status over the past few years. What now seems to be lacking is the requisite legislative framework within which to develop matters further. However, at the time of writing, there are a number of activities that may have a long term influence on shaping future policy. These are the research programme recently commissioned by the DoE regarding data requirements for coastal management, an investigation by the National Audit Office and an inquiry by the House of Commons Environmental Committee. There are also developments within the European Parliament which will be outlined at this Conference (ref. 22).

In addition to these various initiatives recent changes in the policy for provision of grant aid also lead towards encouraging improved shoreline management practices. These are the willingness to consider extended benefit-cost analyses with the inclusion of intangible items and the availability of grant aid for both pre-investment studies and post-construction monitoring. These developments are certainly welcomed within the profession.

Conclusion

This paper has briefly reviewed some historic aspects of the development of coastal engineering design to where it stands at the present day. It can be appreciated that the level of understanding that has evolved over the past half-century has been quite profound and rapid. At the same time it has only been within the last decade or less that the need to manage the shoreline in a much more formal and structured way has been felt to be necessary. This is not to say that various authorities have not always been managing the shoreline within the strict confines of their boundaries, but there have been increasing pressures on the coastal zone resulting in a number of conflicts that did not previously seem so apparent.

The charge that the 'engineers' have carried out coastal works in the past without consideration of the wider issues and sensitivity for the environment has often been made. However, it is not that there has not been an awareness of 'soft option' techniques for a few decades, but the inheritance of extensive lengths of sea walls and groynes should be evident. A greater understanding of coastal processes in relation to coastal defences does not allow the more traditional structures to be immediately replaced as they represent a considerable national asset in investment terms. Given that the provision of grant aid is usually based on the most favourable benefit cost scenario for works that prevent flooding or erosion, this often mitigates towards rehabilitation of the existing structure rather than its replacement with something that would be considered to be more appropriate given a green-field site. Furthermore the general public, particularly in areas that have suffered in the past, perceive solid sea walls as the only acceptable line of defence that will protect them. The concept that some extra sand on the beach or some remote or widely spaced control structures will be as, or more, effective is certainly far from obvious to the layman. In this respect there is clearly a need for education of the public and many efforts are made to discuss the matters in open public debate.

It must also be acknowledged that there are circumstances on which heavy structures are the most appropriate form of coastal defence for use within the coastal zone. In many cases there can be substantial benefits to the local community with only very localised impact on the coastal regime. At the end of the day it is a question of making rational decisions that maintain a sensible balance with respect to the uses in the coastal zone. The prospect of abandoning all coastal works in favour of natural retreat is equally untenable as protection of all areas suffering some sort of periodic inundation from the sea or erosion.

Thus, the design and implementation of engineering works has always been carried out within the framework and constraints provided by the governing bodies at the time. The fact that, in some areas so much property requires protection is largely due to the planning system of the day allowing developments in areas of such high risk. However, it is apparent that major initiatives have been taken in the UK working towards formulation and implementation of integrated shoreline management procedures. Further development into a wider coastal zone management framework should thus be facilitated through open debate as at this conference and other initiatives that are taking place.

References
1. Savile Sir Leopold H.(1940). Presidential Address. J. Instn Civ. Engrs 1, Nov., pp 1-26.
2. Philpott K.L. (1984). Cohesive coastal processes. Engineering Institution of Canada Annual Conference.

SETTING THE SCENE

3. Vernon-Harcourt L.F. (1882). Harbours and estuaries on sandy coasts. Min. Proc. Instn Civ. Engrs, Vol. 70, pp 1-104.
4. Haupt L.M. (1899). The reaction breakwater as applied to the improvement of ocean bars. Trans. Am. Soc. Civ. Engrs, Vol. 42, pp 485-546 and Vol. 43, pp 93-106.
5. Carey A.E. (1904). The sanding of tidal harbours. Min. Proc. Instn Civ. Engrs, Vol. 156, pp 215-302.
6. Carey A.E. (1905). Coastal erosion. Min. Proc. Instn Civ. Engrs, Vol. 156, pp 42-142.
7. Spring Sir F.J.E. (1914). Coastal sand travel near Madras Harbour. Min. Proc. Instn Civ. Engrs, Vol. 194, pp 145-238.
8. Cornick H.F. (1969). Dock and Harbour Engineering, Vol 2 - The Design of Harbours (2nd edn). Charles Griffin, London.
9. Kuhn T.S. (1970). The structure of scientific revolutions (2nd edn). University of Chicago Press, Chicago.
10. Palmer H.R. (1834). Observations on the Motions of Shingle Beaches. Phil. Trans. R. Soc, London, Part I, pp 567-575.
11. Rennie Sir John (1854). The theory, formation and construction of British and foreign harbours.
12. Townend I.H. (1986). Coastal studies to establish suitable coastal management procedures. J. Shoreline Management 2, pp 131-154.
13. Park J.R. (1989). Legislation and policy. In Coastal management. Thomas Telford, London.
14. Townend I.H. (1992). Shoreline management: a question of definition. In Coastal management. Thomas Telford, London.
15. Keay T.B. (1941). Coastal erosion. Dock & Harbour Authority, Oct. (reproduced 1991).
16. Townend I.H. and Fleming C.A. (1991). Beach nourishment and socio economic aspects. Coastal Engineering, Vol. 16, No. 1, Dec.
17. Barrett M.G. (1989). What is coastal management? In Coastal management, Thomas Telford, London.
18. Townend I.H. (1990). Frameworks for shoreline management. PIANC Bulletin No. 71, pp 72-80.
19. Fleming C.A. (1989). The Anglian Sea Defence Management Study. In Coastal management. Thomas Telford, London.
20. NRA (1991). The future of shoreline management. Conference papers.
21. Child M. (1992). The Anglian Management Study. In Coastal management. Thomas Telford, London.
22. Kremer R. (1992). The European Commission policy on coastal zone management. In Coastal management. Thomas Telford, London.

2. Coastal conservation

R. W. G. CARTER, University of Ulster

This Paper deals with the need for, and progress towards, the conservation of the UK coastline. A time when coastal issues, for example oil spills, eutrophication, sea-level rise and habitat loss are becoming more prominent, particularly in the eyes of the public, is a good time to reflect on what is being conserved and why. Axiomatically the next stage is to examine how effective past conservation efforts have been, and what, if anything, might be done to improve them.

Conservation

Conservation is the equitable maintenance of the environment for the benefit of its inhabitants. In coastal terms, this means that the shoreline remains in good working order, so that all its users can combine to make use of its resources. Thus conservation has many facets, so that it may be hard to balance supply of the resource against demand. Certainly this is the dilemma facing shoreline managers as the coast becomes more and more popular. Stressed environments are harder to control, and conservation objectives are often limited to one or, at best, a few. Over the last century, the development of the UK shoreline has proceeded apace, with industrial, commercial, residential and recreation activities all competing for space. During this expansion period many natural shoreline environments have been damaged or destroyed, including wetlands, dunes, beaches and cliffs. Progressive loss of habitat has often resulted in a decline in both species diversity and numbers, a situation which, paradoxically, favours the promotion of conservation. However, it is important to recognise that conservation is not preservation, or selective exploitation or sustainable development; it is concerned directly with the health and well-being of the environment, in this case the UK coastline.

Conservation may be tackled from a number of standpoints, through the mediums of education, legislation or altruism and by the mechanisms of land ownership, accessibility, enforcement and accountability. All combinations of these can be recognised as part of the UK coastal milieux. However, good conservation practice can only be based on two things

(*a*) understanding how the system (in this case how the coast) works
(*b*) within an established framework of effective, far-sighted management, embracing policy and practice.

SETTING THE SCENE

How the UK coast works

The coast is a dynamic environment involving constant adjustments between land, water and air. The coast of the British Isles occupies a position on the downwind margin of a major ocean basin. The coastal boundary acts to redistribute, absorb and reflect large amounts (of the order of 10^9 J/m/year) of wave energy. Swell wave energy penetrates much of the Irish Sea, largely through St. George's Channel, and the North Sea, entering mainly from the north between Shetland and Norway. The epicontinental seas respond readily to the frequent west to east moving cyclones. Also, tidal energy is dissipated along the coast, especially in the English Channel, Bristol Channel, the Firths of Clyde and Forth and the Wash. Between waves and tides there is a constant adjustment of the sediment budget.

The major recent geological influence on the British coast has been the Pleistocene glaciations which have both provided a source of sediment and controlled the sea-level trends over the last 15000 years. The coast has received sediment from three sources, the shelf, rivers and cliffs. Nationally, the former is most important, with the others assuming local importance. By and large the supply of fresh sediment to the coast has declined exponentially since the Ice Age, so that the British coast has switched from one of abundance to one of scarcity. This scarcity has been compounded over the last two centuries through the building of seawalls, jetties, groynes, the regulation of rivers and the removal of sediment.

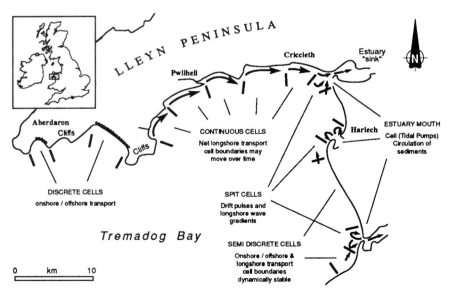

Fig. 1. *Tremadog Bay in North Wales contains at least three types of coastal cell*

The coast of Britain may be divided up into a series of cells (Fig. 1), within which there is a basic recirculation of materials. Cells are "driven" by incident waves, tides and river/estuary currents. Cell boundaries move in time and space, so that sediment may be exchanged or subsumed within adjacent or overlying units. Where sediment is locally abundant, cell structure is often immature, as processes fail to find a balance with supply. Along the East Anglia coast, inputs of 10^3 to 10^4 m^3/year from cliffs maintain a cell-independent drift over several tens of kilometres. Where sediment is scarce the cell structure is often strongly developed, with discrete units and little sediment exchange alongshore. In places the maturity of a cell is defined further by decoupling or partitioning of sediments by size or shape, to form graded beaches. Cell structure is often hierarchical, with one cell embedded within another, perhaps indicating a different time scale for development (ref. 1).

The relationship between coastal processes and morphology is dynamic. Constant variation in incident processes requires a range of energy-dispersive mechanisms, both across and along the shore. Recognition of the importance of secondary fluid motions (from sub-harmonic to infragravity) has led to the identification by morphodynamic coding of shorelines. One extreme is represented by vertical cliffs causing complete reflection, the other by near horizontal slopes in which frictional dissipation removes all incident energy.

Combining the ideas of cell development with morphodynamics allows consideration of meso-scale coastal evolution (decades to centuries, 10^1 to 10^2 km). In many cases there will be domain shifts (progressive or sudden) as control variables change. Much of the coastline of Britain is still evolving slowly, with sediment virement taking place, with material showing a net movement downdrift towards relatively low energy sinks such as the southern North Sea Liverpool Bay, the inner Bristol Channel, the Firths of Clyde and Forth and Solway. Perturbations to this pattern are often provided by river mouths or estuaries which have the capacity to capture and circulate sediment. River mouth efflux sets up a sediment pump which embraces not only submarine bars and the shoreline, but also coastal dunes. Much coastal sediment in the British Isles is stored at those locations. Estuaries tend to accumulate sediment, both from landward and seaward sides. Over time the slow infilling of estuaries may lead to hydraulic inefficiency and closure of their mouths.

The ecology of the coastline has developed within the framework provided by the physical environment, and in this sense organisms are vulnerable to any changes that might occur. There is considerable efficiency within coastal communities, as nutrients and minerals are often recycled and reused. For example, productivity of primary producers in the nearshore zone depends on the wave pumping mechanism which favours accumula-

SETTING THE SCENE

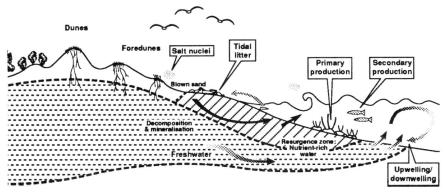

Fig. 2. The shoreline performs an important ecological role in terms of decomposition, mineralisation and nutrient cycling

tion of tidal litter, followed by bacterial decomposition on and within the beach, and resurgence of nutrient-rich water into the nearshore zone (Fig. 2). Storm activity helps this process. Furthermore, a more complex pathway exists between nutrient production, breaking waves, dune development and freshwater efflux, which stimulates more inshore productivity.

In fact the coastal ecology is linked through a complex series of interactions. Estuaries and rock coasts are especially important (ref. 2) as they sustain highly productive ecosystems (often exceeding 500 gC/year/m^2), with the potential to export nutrients to adjoining areas. Thus a basin like the Irish Sea is sustained by a series of marginal estuaries, including the Ribble, Solway, Mersey and Conway on the east side and Belfast, Strangford and Carlingford Loughs on the west (Fig. 3). The chain of estuaries and coastal wetlands assumes importance in maintaining fish and bird populations, and represents a series of feeding stops for migratory birds. Thus the impact of man on estuaries which have been reclaimed over the past 600 years has had severe repercussions on the coast (refs 2 and 3). Index species like dolphins have declined dramatically in the last century.

Impact of man

Virtually none of the British coast is natural. In recent centuries almost all the coastline has been subject to changes of one kind or another. Coastal resources in terms of fish, birds, vegetation and sediments have all been exploited, and in many cases lost altogether. Attempts to control coastal processes probably started in the early Middle Ages when Marram Grass *Ammophila arenaria* was first planted to fix mobile dunes.

One of the most pernicious problems is that of coastal erosion and numerous attempts have been made to halt recession. (Yet it is important to

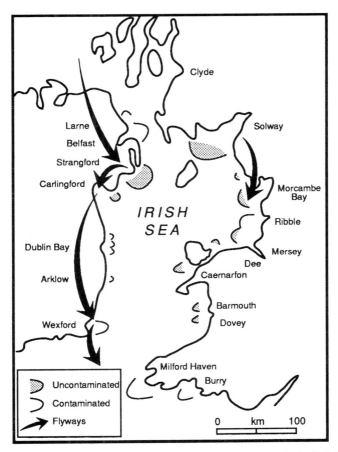

Fig. 3. Conservation of the Irish Sea estuaries is vital to maintain both fish and bird populations

remember that most erosion problems are due to man getting too close to the sea rather than the sea getting too close to man.) Measures to control erosion have traditionally involved the construction of seawalls and groynes to arrest mobile sediment. This may cause downdrift starvation and initiate erosion elsewhere. More subtle effects include the curtailing of the natural beach/cliff sediment exchange (especially on dune coasts) and the alteration of the shoreline water table . Both impacts may encourage the lowering and disappearance of the beach. This cause-effect relationship may take years to become apparent, maybe a century or more, although the net result is to impair the ability of the natural system to respond to environmental change (ref. 4).

SETTING THE SCENE

Manifestations of human impact are especially obvious in and around estuaries. The reclamation of estuary wetland reduces the tidal prism and the productivity of the ecosystem. The former leads to a shrinkage of tidal bedforms, adjustments to the outer estuary shoreline and ultimately may herald the onset of erosion some distance from the original reclamation site (Fig. 4). The loss of productivity has a clear impact on the food chain and

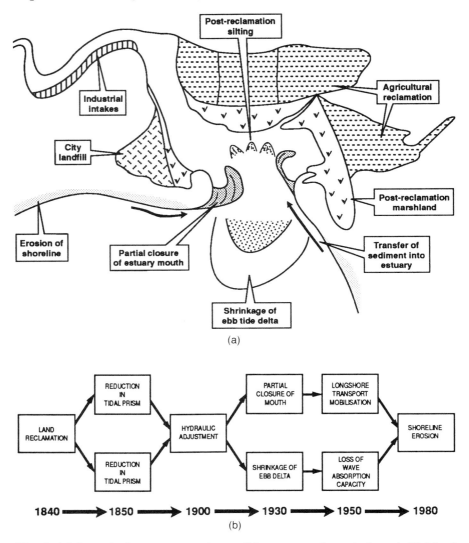

Fig. 4. (a) Impact of man on an estuary; (b) sequence of events from initial land reclamation to shoreline erosion (the time scale is hypothetical, but realistic)

species which cannot adapt or migrate elsewhere may die out, leading to a reduction in biodiversity. Further complications may arise through the leaching of sulphates and aluminium silicates from drained land, leading to acidification of coastal waters.

Very few British coastal sites remain unaffected by man. An example illustrates this point. The coast of southwest Scotland comprises a series of northward-opening bays separated by rocky outcrops. Around Ayr (Fig. 5), the coast was first altered in the eighteenth century when the port was developed. Since that time sediment starvation, plus increasing urbanisation, industrial and recreational use of the shore, has led to erosion and aesthetic problems, particularly around Troon. Various tactics have been employed to halt erosion, but the quality of the coastal environment has suffered further, and the natural capacity of the system to achieve balance has been impaired.

In this and many other examples the impact of man has been slow, but progressive. The roots of many of the present problems must be sought in former actions, stretching back two centuries although a prolonged sequence of changes is often dealt with incrementally, usually as a response to events. It is often difficult for local people to perceive such long term trends and therefore to associate effect with cause. Moreover such time periods fall outside any legal statute of limitations. One problem often creates another and the standard UK response has been to react to each, often by constructing further sea defences. Bournemouth in Christchurch Bay has received 47 separate treatments since 1900 (ref. 5). A sequence of reactive solutions

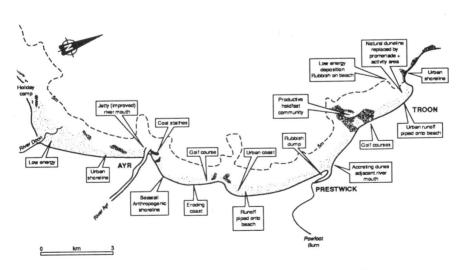

Fig. 5. *The use and conservation problems of the Ayr/Troon coastline*

SETTING THE SCENE

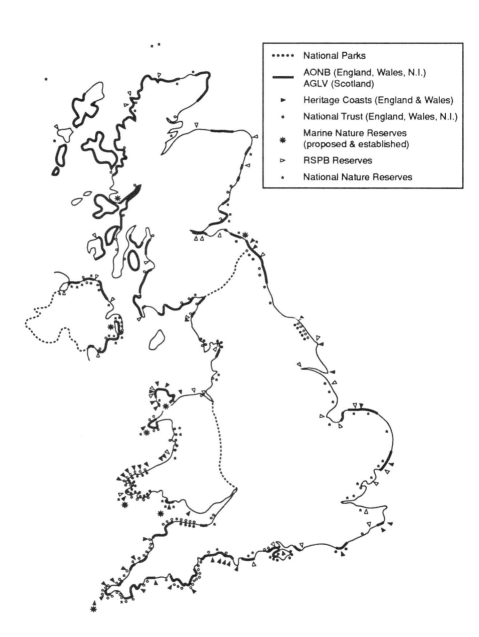

Fig. 6. Coastal conservation designations in the United Kingdom

constitutes a policy trap, difficult to escape from and almost always an expensive way to proceed.

Conserving the UK coast

A conservation policy for the UK coast has never been explicitly adopted (ref. 6), although the rudiments of conservation policy date back to the 1930s and 1940s some of which became part of the 1944 and 1947 land use planning legislation. The current praxis stems from the National Parks Commission initiative in the 1960s which led directly to the establishment of Heritage Coasts in 1974, and the National Trust's Enterprise Nature project, started in 1965 (Fig. 6). As well as these major schemes a number of local authorities (e.g. Sefton) and voluntary bodies (e.g. RSPB) also conserve coastland although objectives tend to vary from group to group and site to site (refs 7 and 8). It may be argued, quite reasonably, that long stretches of the UK coast are conserved in as much as they fall within one or other of the statutory designations used in planning [Areas of Outstanding Natural Beauty (AONB) or Areas of Great Landscape Value (AGLV)] (ref. 9). However, these designations may attract development as they are seen as desirable places to live or visit. In addition, small areas of coast are protected within Sites of Special Scientific Interest (SSSIs) under the auspices of the Countryside and Wildlife Act.

The Heritage Coast concept aims to protect undeveloped coast through a combination of management and legislation, but with emphasis on the former, favouring consensus between coast users (ref. 10). Altogether there are now 43 Heritage coasts in England and Wales, covering about 33% of the coast. While directed by the Countryside Commission, via the Heritage Coast Forum (established in 1987), day-to-day management is vested in local wardens and officers who liaise with coastal authorities and landowners. Among private owners of the coast, the National Trust is by far the most influential in terms of conservation. The Enterprise Neptune appeal launched in 1965 was aimed specifically at protecting scenically-valuable coast. By 1989, almost 847 km had been obtained, much of it in southwest England, Wales and Northern Ireland. As custodian of such scenic resources the National Trust's role is beginning to change from one of conservation through exclusion to a more market-orientated approach, which collects entrance fees, provides access and information and sells souvenirs.

Conservation issues

Conservation of the coast often arises through enlightenment that coastal resources have a hedonic value, which is not easily replaced if lost. Several issues in particular have heightened this awareness, especially the destruction of coastal wetlands, decline in both number and diversity of seabirds, and damage attributed to recreation visitors. While measurement of losses in environmental quality is not often subjective, many coastlines have experienced marked changes over the last decades. The intrusion of first caravan sites into dunes and more recently marinas into almost every small estuary and harbour in southern Britain, both correlate with loss of environmental quality. The response of many entrenched conservationists has been to stiffen access controls, create diversionary routes, land-use zoning, increase policing by wardens and to develop information/interpretation centres.

Paradoxically increasing pressure has probably increased ecological diversity at the coast, as nutrient enrichment may boost rocky shore productivity and attract new species, trampling of dunes may provide niches for exotic plants and migrant invertebrates and tidal litter of human origin presents new opportunities for insects. In many areas, somewhat oligotrophic communities have been enriched and diversified by newcomers in the wake of human disturbance. Many species have been introduced, *Spartina anglica* has invaded much of the intertidal silt around the coasts of England and Wales following its inadvertent introduction (as a genetic cross) in 1870 (ref. 11). Much of the *Spartina* was planted and spread naturally only within bays. Rabbits were introduced to dunes in the Norman period, and later dunes were developed as commercial warrens, in places supporting thousands of animals per hectare for meat and fur. Other invaders include Rhododendron, Sea Buckthorn and Sycamore on dunes and Pacific Oysters in the intertidal zone. In all cases, these exotic species have caused management problems, destroying the natural ecological balance. However, some like Sea Buckthorn have advantages in that they may support alternative ecosystems.

Long-term trends are also evident. Vegetation monitoring in the Netherlands has pointed to increasing acidification of dunelands especially on those that are calcium-poor (ref. 12), where both vegetation and soils are changing towards low pH tolerances, with inevitable loss of diversity. In many coastal areas, excess nutrients, especially from agricultural waste and domestic sewage are leading to eutrophication. Recent algal blooms in shallow water in the Baltic, southern France and the Adriatic have led to deoxygenation and fish-kills (ref. 13). However, it is evident the coast is under severe ecological pressure; in 1985 the EC reported (ref. 14) that one third of European coastal dunes had been destroyed and up to half the salt marshes irrevocably altered by man in the last century.

As well as ecological issues, there are many geomorphological problems, particularly associated with engineering structures disrupting natural sediment transport pathways. Inappropriate construction of groynes and seawalls may lead to sediment diversion offshore, beyond the normal beach/dune/nearshore exchange. The removal of backshore dunes may damage the natural water table and allow penetration of saline water inland.

Coastal dunes are especially vulnerable to human impact, as they are often viewed as wasteland suitable only for development. However, dunes act to shelter inland areas from high winds and salt burn, buffer the shore against wave attack and maintain the coastal water table. Removal of dunes may impair all three functions, in places eliminating them altogether. Dunes are sensitive environments. The "healthiest" dunes are those with a degree of instability which fosters natural diversity. Totally stabilised landscapes are vulnerable to damage or die-back and are unable to recover. Totally unstable dunes may engulf inland areas and encourage coastal recession.

Another vulnerable coastal landform is the gravel beach or ridge which is common in Britain. These features have been badly damaged by coast protection works (especially groynes) so that they have begun to break down, prompting further protection, as at Hurst Castle spit on the Solent (ref. 15). Gravel beach and ridge ecology has also suffered. At Dungeness abstraction of fresh water, plus gravel-pit development has led to salt water intrusion affecting the unique vegetation of the site (ref. 16).

Until recently aquaculture had had no great impact on the coast although the expansion of fish farms in western Britain and Ireland has led to expression of fears particularly in relation to genetic dilution of wildstock and release of pesticides. Clearly fish-farming is undergoing an acceptance phase in relation to conservation, and many remain sceptical despite the fledgling industry's protestations of eco-friendliness (ref. 16).

Current and future trends in coastal conservation

Over the past decade, political changes have influenced many conservation issues, not least the rise in green consciousness, the related interest in the media (especially in this context of the spectre of sea-level change), the trend towards increasing internationalisation, the emergence of free-market doctrines and rightwing hegemony.

The green movement emphasises the use of environmentally-benign or beneficial techniques (ref. 18) especially "soft" engineering solutions to problems such as coastal erosion (ref. 19). These techniques such as vegetation planting, the use of low-cost materials and beach nourishment are also finding increasing acceptance by engineers, although full appreciation of the sensitivity and application of such methods is still some way away (ref. 20). For instance the planting and cultivation of salt-tolerant vegetation requires considerable skill, especially in terms of maintenance before establishment. Many projects fail at this juncture due to lack of sufficient aftercare.

Media interest in coastal issues is useful in raising awareness. The well-known attention cycle requires the media as a catalyst for political change,

SETTING THE SCENE

so that acceptable political, legislative or financial assistance may be secured. In the USA the geologists versus engineers debate of the early 1980s (ref. 20), in which the value of structural solutions for beach erosion was criticised, resulted in media interest and ultimately the banning in 1986 by five east coast States of seawalls and groynes in favour of softer techniques.

Increasing internationalism sees the advent of European coastal policies, aimed at conservation of what the CEC sees as a major resource, through a number of funded programmes including MAST, EPOCH, STEP and ENIVREG. There is also the Birds Directive (79/409/EC), Bathing Waters (76/105/EC) and the new Habitats Directive, all of which have a strong element of coastal conservation. It seems that future action could require member states to develop Coastal Zone Management Plans to ensure integrated development is carried out in an economically efficient and environmentally sensitive manner. This would extend the EIA Directive (85/337/ EEC) provisions to an area rather than a specific type of development. This plan is redolent of the US Coastal Zone Management Act (CZMA) of 1972 which was designed for very similar objectives. Despite criticism, the CZMA promoted a wide range of research and discussion, and has, on balance, led to tangible improvements in conservation along US shorelines (ref. 22).

The impact of "Thatcherism" on coasts is far more subtle (ref. 23), but includes direct changes in legislation, privatisation (especially of Water Boards, Hydraulics Research and such small concerns as the Poole Harbour Board), the encouragement of the environmental services industry, refinancing the Countryside Commission and altering the structure of the Nature Conservancy Council. The pros and cons of these changes have yet to be fully assessed, but the net result might be a more realistic economic assessment of grant aid for coastal projects.

The future requirements for coastal conservation must be based on scientific understanding, especially if both economic and environmental objectives are to be met. The last 20 years in the USA have seen a plethora of mechanisms developed to help conserve coastlines. This includes, in no particular order, trading of coastal rights, zoning of shorelines, tax differentials, habitat protection bills, changing of access rights, development credits, environmental mitigation policies, environmental audits, pollution clean-up penalties, bubble policies, citizens' bond issues, withdrawal of insurance, compulsory purchase, imposition of non-transferable property rights, extension of statutes of limitations, culpability clauses and environmental class actions. All these measures aim at cajoling coastal users to adopt more environmentally-sound approaches, by restricting activities and development. The State of California now insists that developers of coastal wetland create up to four times as much new wetland to replace the area destroyed. Florida has legally-enforceable set-back lines to stop development too close to the shore. This menu of possible approaches is neither wholly sensible nor desirable (nor exhaustive) but simply serves to show what might be applied elsewhere. Several of these tactics have been tried in Europe, but the key to their success is very much in their enforcement.

The type of enforcement varies according to the conservation requirements (Table 1), and may range from severe (in terms of legal prohibition) to moderate (for example with public agreements - usually achieved through education). The legal system in the UK relating to conservation is relatively unwieldy, especially compared to the US where numerous avenues are open to conservationists to challenge deleterious practices, including "biotic rights", "no significant degradation", "fairness principles", "public trust" and "*qui tam* initiatives" (based on informers). All these methods have been employed in US courts to uphold environmental standards.

As well as improving the legal framework for coastal conservation and placing it within the context of a strategic coastal management plan, two other perhaps more radical steps could be taken. These include

(*a*) a greater emphasis on preventive rather than protective measures
(*b*) a switch from capital-intensive to recurrent-intensive expenditure.

Table 1. Conservation enforcement and compliance

	Type	Example
Expressions of law	International laws CEC Directives Acts of Parliament By-laws Statutory rules and regulations	Maritime boundaries Bathing water Coast erosion Access Water quality
Licences	Standards Requirements Regulatory conditions Consents	Waste disposal Dumping at sea Sea defence Sediment extraction
Voluntary agreements	Resolutions Recommendations Codes of Practice Management agreements Participatory consensus	SSSI designations Heritage coasts Sports activities Nature conservation Leisure activities

SETTING THE SCENE

The first implies more foresight in order to circumvent problems before they arise. Too much coastal conservation in Britain arises as a consequence of actions, rather than as a precursor of them. Part of this is a tendency for UK coastal development to be speculation-led rather than demand-led. Preventative conservation would also emphasise the dynamics of the coastal system, rather than its statics. For example conservation of geologically important coastal cliffs (ref. 24) (which require continued erosion) could be best achieved through exploring the requirements of the long term regional sediment budget, rather than trying to tackle the problem as site-specific. A national preventive strategy for coastal conservation could avoid the pitfalls of unnecessary protective works which have so disrupted the UK coastline over the past 150 years.

To achieve this aim, there will need to be a switch from capital to recurrent expenditure which is more beneficial to management. Present UK Government Policy (as with all exchequers led by finite-term democratic parliaments) favours capital allocation (euphemistically called 'investment') over annual recurrent expenditure. In this context, much coastal conservation could be achieved if the budget favoured longer term, lower capital cost projects, covering research, monitoring, education and wardening. The recent House of Commons Select Committee Inquiry into coastal management in the UK may signpost another way ahead.

Conclusions

Coastal conservation in the British Isles is a hit and miss affair. Over the past century much conservation has been achieved through a mixture of luck, individual skill and institutional altruism. The success of the National Trust's Enterprise Neptune scheme and the Countryside Commission's Heritage Coasts tends to mask a more general impoverishment which the policies of the last decade may have exacerbated.

Good coastal conservation can only be built on a proper understanding of coastal processes, integrated across a range of time scales. There are numerous techniques available, but there should be more proactive approach to problem prevention rather than solving, a shift of resources towards longer term management and a breakaway from conventional or off-the-shelf solutions which tend to retard progress. Policy, planning and practice should be scrutinised with the intention of finding a better way towards effective conservation of one of the Nation's most valuable resources.

References

1. CARTER R.W.G. Coastal environments. Academic Press, London, 1988.

2. DAVIDSON N.G. et al. Nature conservation and estuaries in Great Britain. Nature Conservancy Council, Peterborough, 1991.
3. PRATER A.J. Estuary birds of Britain and Ireland. Poyser, Stafford, 1981.
4. ORFORD J.D. Alternative interpretations of man-induced shoreline erosion in Rosslare Bay, southeast Ireland. Transactions of the Institute of British Geographers, New Series, vol. 13, 65-78, 1988.
5. LELLIOTT R.E.L. Evolution of Bournemouth defences. In Coastal Management, Thomas Telford, London, 263-277, 1989.
6. GUBBAY S. A future for the coast? Proposals for a UK Coastal Zone Management Plan. Marine Conservation Society, Ross-on-Wye, 1990.
7. HOUSTON J. and JONES C.R. The Sefton Coast Management Scheme: project and process. Coastal Management, vol. 15, 267-297.
8. HOUSTON J. and JONES C.R. Planning and management of the coastal heritage. Sefton Metropolitan Borough Council, Southport, 1990.
9. HALLIDAY J. Coastal planning in England and Wales. Ocean and Shoreline Management, Vol. 11, 211-230, 1988.
10. BRADBEER J. Heritage coasts in England and Wales. In Land, water and sky: European environmental planning (G.J. Ashworth and P.T. Kivell, eds.) Geopers, Groningen, 121-136, 1989.
11. HUBBARD J.C.E. and STEBBINGS R.E. Distribution, dates of origin and acreage of *Spartina townsendii* (s-l) marshes in Great Britain. Proceedings of the Botanical Society of the British Isles, vol. 7, 1-7, 1967.
12. KOERSELMAN W. The nature of nutrient limitation in dune slacks. In Coastal Dunes: Geomorphology, Ecology and Management (R.W.G. Carter et al., eds.) Balkema, Rotterdam.
13. FUNEN COUNTY COUNCIL. Eutrophication of coastal waters. Danish Department of Technology and Environment, 1991.
14. GEHU J.M. European dune and shoreline vegetation. European Committee for Conservation, Strasbourg, 1985.
15. NICHOLLS R.J. Poole and Christchurch Bay in the coastal sedimentary environments of Southern England, South Wales and Southeast Ireland (R.W.G. Carter et al., eds.). BSRG, Cambridge, 6-31, 1991.
16. FERRY B. and WATERS S. Dungeness. Ecology and conservation. Nature Conservancy Council, Peterborough, 1985.
17. SCOTTISH WILDLIFE AND COUNTRYSIDE LINK. Marine salmon farming in Scotland: a review. SWCL, Perth, 1990.
18. PORRITT J. Seeing Green. Blackwells, Oxford, 1984.
19. BROOKES A. Coastlands. British Trust for Conservation Volunteers, Aldershot, 1979.
20. AANEN P. et al. Nature engineering and civil engineering works. Pudoc Wageningen, 1991.
21. PILKEY O.H. Geologists, engineers and a rising sea-level. Northeastern Geology, vol. 3, 150-158, 1981.

SETTING THE SCENE

22. ELIPOULOS P.A. Environmental Reporter Monograph, vol. 30, 1-48, 1982.
23. CARTER R.W.G. Coastal management in the United Kingdom, 1980 to 2000. Proceedings Coastal Management Symposium, Nantes (in press).
24. HYDRAULICS RESEARCH. A summary guide to the selection of coast protection works for geological sites of Special Scientific Interest. Wallingford, 1991.

3. Setting the scene: the planning dimension

G. A. D. KING, Environmental Planning Consultant and Chairman, National Coasts and Estuaries Advisory Group

Introduction

A lot of water has passed under the bridge - literally and metaphorically - since the conference in 1989. Rereading those papers recently, I was impressed with the breadth of expertise and foresight already evident in the face of the emerging problems. Climatic change, environmental issues, the Anglian Sea Defence Study, the Hampshire planning approach, case studies such as Carmarthen Bay - papers on these subjects all showed recognition of the important inter-disciplinary context in which we now know we have to work. My task today is to set the planning scene - a field where much has happened over the last three years, prompted also by an important Planning Symposium, again held in 1989, in Southport, which concluded with an Agenda for Action comprising the following points.

(*a*) There had to be an update of National Guidance beyond circular 12/72 on 'The Planning of the Undeveloped Coast', which introduced the Heritage Coast concept, rather under estimated the conservation value of 'mud', and thus neglected the importance of estuaries both in wildlife and scenic terms.

(*b*) Improved co-ordination of coastal information and action was necessary at national, regional and local levels.

(*c*) The profile of coastal issues should continue to be raised.

Pressure for change

If progress has been slower on (*a*) and (*b*) above than many would have wished, on issue (*c*) the future of our coastline has certainly emerged as one of the leading political concerns in the green debate. As one of my colleagues has written: "Nowhere more than in the UK is the coast so much a part of the national psyche; nowhere is the planning of the coast more fragmented. This is a green issue if ever there was one ... vital aspects of nature conservation, complex issues like barrages (here he might have included coast protection), the recreational use of the coast - a whole pot pourri of prime political and planning hot potatoes".

SETTING THE SCENE

Indeed the Government has been besieged by a welter of reports and advice upon the subject, further conferences have been held, both national and European (including Prince Charles's foray into the North Sea), associations have sprung up (like my own group, engineering cells, the Cardigan Bay group, and the French-based Les Esturiales), while nature, at places like Towyn on the north Welsh coast, have helped focus the politicians' minds in ways that prophecy and speculation never can.

So we have had excellent reports and studies such as the RSPB's "Turning the Tide" (followed more recently by a pre-election manifesto on how to save the World - if the birds prosper so shall we!), and the Marine Conservation Society's "A Future for the Coast?", a well-argued report which I believe served to provoke the Keep Britain Tidy Group, supported by the Duke of Edinburgh, into holding their conference last summer on "The Coastline - A Wasting and Wasted Asset". More recently the Marine Conservation Society has issued a Discussion Paper introducing the idea of a Coastal Zone Management Unit.

Essentially what the pressure groups are so passionate about are the contingent threats to wildlife, ecology, and natural beauty, now established by survey and research. The RSPB report indicates that 80% of estuaries appeared to be under some type of threat, with 25% at risk of permanent damage from an array of sources, many of them not amenable to planning control. New guidelines, powers and machinery are called for, in a similar view to the recommendations contained in the Marine Conservation Society's report.

The planning response

In a response for the County Planning Officers' Society on these documents, based on my own survey of maritime county planning authorities, I concluded that good practice in the planning and management of estuaries was patchy or neglected, that Heritage Coast designations frequently did not include estuaries, and that despite comprehensive coverage of strategic policies in structure plans these were not always successfully translated into programmes of action. Indeed, the RSPB's own Planscan report, based on research into the clarity and effectiveness of the wording of policies in Structure Plans, also confirmed my own misgivings. Estuaries were also especially vulnerable as they often provided the traditional boundary demarcations between local authorities. What was more disturbing was the fragmentary fashion in which the Coastline was treated in the Government's then recently issued Green Paper on "Our Common Inheritance". As a result the County Planning Officers' Society expressed their concern through the Association of County Councils to the Government and also established the Advisory Group on Coasts and Estuaries of which I am Chairman, and which includes planning officers from the Association of District Councils and Association of Metropolitan Authorities. Our object is to disseminate information, promote good practice, and provide observations on the many issues of current concern.

Given the long-established and well-respected basis of the British Town and County Planning system, and our record of achievement in protecting many parts of the coastline through existing plan-making and development control procedures, our concern may seem surprising. But remember, planning control only extends to the low water tidal definition, and as county planning authorities we had run the gauntlet through much of the 1980s of a Government determined to abolish structure plans, having already introduced a 'presumption in favour' of proposed development (except for AONBs and certain other defined areas). At one stage Green Belts too were very much at risk. In this atmosphere - subsequently reversed I am glad to say, or at least, as far as Structure Plans are concerned, deferred until smaller scale unitary development plans are produced, following local government reorganisation - the proposed withdrawal of subject plans, of which many good examples were devoted to coastal conservation, such as Suffolk Heritage Coast, suggested a Government less than determined to achieve a holistic approach to planning the coast.

The Government approach

Nevertheless the pressure on the Government continued, and the basis of our concern was also confirmed by the final publication of the former Nature Conservancy Council's study of Estuaries. At a two-day, multi-disciplinary conference on the Coast, organised by the Royal Society of Arts, this and other contributions made the same point concerning the indeterminacy over what to do. It was a relief therefore to learn of the DoE's decision to commission a wide-ranging review of coastal affairs as, and I quote, "the coastal zone is a matter of increasing concern to many organisations". Research was to have a two-fold purpose: to secure a basis for Planning Policy Guidance and also on the use of earth science information in coastal zone decision making. The study would review the scope and interaction of legislation, responsibilities of different bodies, the effectiveness of current mechanisms for handling conflicts, and the effect of rising sea-levels. Although very encouraging, we have subsequently learnt from Government officials that the forthcoming PPG will not involve any policy change, or changes, to powers or institutions, although it might present existing policy in a new way.

This approach is confirmed in the Memorandum on Coastal Policy prepared on behalf of Government ministries for the House of Commons Environment Committee Review of Coastal Zone Protection and Planning,

SETTING THE SCENE

in which we learn the Government has not so far seen convincing evidence to support the claims of confusion over the multiplicity of agencies and does not therefore believe a single agency is either feasible or desirable. In the Government's view, appropriate mechanisms are in place to reconcile most of such conflicts. Most of the evidence I have seen suggests this is not the case, although it must be admitted that so much attention to the issues involved has resulted in a greater sprit of co-operative working by those involved. But much remains to be done.

National Coast and Estuaries Advisory Group

At this point, it is perhaps appropriate for me to outline briefly the views of my own group of professional planners, on behalf of the Local Authorities, in our submission to evidence to the Committee of Inquiry. We summarised the main inadequacies as follows

(a) lack of co-ordination nationally/regionally
(b) poor liaison over specific problems
(c) mismatch between existing legislation and needs
(d) inter-user conflicts
(e) estuaries neglected
(f) narrowly defined port and harbour responsibilities
(g) lack of knowledge on cumulative impact of coastal activities, including off-shore dredging
(h) challenge of sea-level rise for coastal protection
(i) continuing pollution
(j) poor progress on marine conservation
(k) local government reorganisation.

In many respects our findings are not dissimilar from the conclusions of the European Workshop held in Dorset in April 1991.

A great deal of innovative work has been undertaken by authorities within the complexities of the existing situation, and many examples of successful co-operation exist, although recommendations frequently depend upon the goodwill of those involved. Many issues can be tackled within the present system, given a clear lead by Government. Nevertheless we welcome the searching inquiry into inadequacies of the existing system, and in our submission deal in some detail with the most pressing problems. In this paper I touch upon some of the main areas of concern.

The main issues

First is the question of the extension of statutory planning control over inshore waters. In Poole Bay, in connection with the Wytch Farm oil reserves, this is being achieved through Parliamentary agreement to the inclusion of a provision in a Private Bill - a cumbersome way of proceeding. Arbitrary lines to delineate local authority boundaries adopted by the Ordnance Survey in the nineteenth century again are unhelpful. Domestic legislation to overcome such problems unfortunately exempt such developments from the terms of the EC Environment Assessment Directive, although there are proposals to change this.

SSSI boundaries drawn along the mean low water mark have also caused problems, in so far as they fail to give any protection to those areas below mean low water mark even though they may be of equal scientific value. Even where designation occurs, especially on estuaries, as English Nature points out, damage still occurs - in the three years between April 1986 and March 1989, 146 SSSIs were permanently or lastingly damaged. The complexities of currently complying with the requirements of the Ramsar Convention will I am sure be dealt with in the conservation workshops.

We are also concerned with the narrow remit of port and harbour authorities, as their role grows in importance as water recreational activities continue to grow. Many are hindered by out-of-date legislation which inadequately deals with conservation and recreation issues. The exemptions from planning control which ports and harbours currently enjoy under the Town and County Planning General Development Order are unacceptably wide-ranging.

The unique position of the Crown Estates Commissioners (CEC) has been criticised in many quarters. The CEC, which owns much of the foreshore and the sea-bed on behalf of the Crown, has considerable powers over the use to which the sea-bed is put, but few of the responsibilities for environmental protection which should be associated with such powers. In effect, the CEC is judge and jury on many developments in the coastal zone.

The number of organisations with powers or responsibilities in the coastal zone is over thirty. What might be best done about this, I hope, will emerge from the Government response to the research commissioned by the DoE, and from the deliberations of the House of Commons Environment Committee. Many solutions have been put forward ranging from modest tinkering, and limited rationalisation, through to the radical reorganisation of Government Departments. Needless to say no-one wishes to relinquish their powers. The RSPB for their part have called for a National Coastal Authority to co-ordinate strategic regional planning in the coastal zone, responsible for planning and co-ordinating matters affecting the sea surface as well as the sea-bed. They say the NRA should be strongly considered for the role, and that the artificial split of 'planning' functions between landward and seaward elements should be replaced by a holistic approach managed by local authorities where planning jurisdiction should be extended 'out to sea' to cover marine developments.

SETTING THE SCENE

In a similar vein, The Marine Conservation Society has recently called for a CZM Unit to provide advice on rationalizing legislation, providing a framework for plans and the funding of programmes. This is easier said than done, particularly in the light of proposals for an Environmental Protection Agency, but such ideas do help to focus attention on the problem.

For our part, as local authority advisers, like many others we have called for improved national guidance involving the regional dimension, the establishment of a standing forum - embracing all interested parties - for each relevant stretch of coast and estuary, with the designation of a 'lead agency' to convene it, charged with taking a strategic and longer term view of coastal protection, planning and management within its area. Policies emerging from these management plans would provide guidance to authorities preparing structure and local plans. Concerning territorial jurisdiction we recommend that the Secretary of State should have powers to define the appropriate boundaries according to local circumstance.

Strategic planning of the coastline

In an overview of this kind, space does not permit the detailed exploration of what might best be done. I wish therefore to highlight a particular area of concern which seems to be rather confused. In their evidence on Coastal Policy the DoE have confirmed that the Government is preparing a PPG Note on coastal planning (albeit with the limitations I have mentioned); that it will monitor voluntary coastal zone management plans; and that it will conduct a periodic review of coast protection and flood defence policies. However, in paragraph 117, in their section on Coast Defence, the DoE affirm that MAFF and the Welsh Office have both recently initiated a coastal defence forum through which groups can liaise on technical matters of common interest, and which, they say, "provides a new mechanism for encouraging strategic planning of coastal defence on a coastline basis". This of course is commendable, but what causes me concern is - taking Wales as an example - that there we also have a Coastal Strategy Working Party formed under the aegis of the Strategic Planning Guidance exercise for Wales (known as Regional Guidance in England). The one group appears to be engineering based, related to drawing up coastal programmes relating to coast protection and sea defence, albeit with immense environmental consequences, while the other is a longer-term planning and essentially environmental exercise. But when and how will the twain meet?

In the not too distant future, after local government reorganisation, Structure Plans will be replaced by district-wide Local Plans, which will probably be too broad-brush to include detailed assessments and policy guidance for their individual stretch of coast. From returns to our national survey on plan preparation, it is clear that not all district or county planning officers were aware of the engineering cell studies taking place on their stretch of coastline. The obvious danger concerning consultation between key parties is that there may be too little too late. In the context of sea level rise, the prospect of storminess, of soft engineering solutions and set back proposals, it seems to me to be crucial that talking, and understanding, should begin now. And if we do not sort it out, then the Government should. Indeed, as a measure of our concern on the newly formed Heritage Coast Forum Steering Group for Wales, we have appointed a Consultant Coastal Engineer to our group. This is why I so much welcome the chance to present this Paper and to take part in the detailed discussion on these crucial issues of practice. Strategic planning of the coastline is not the same exercise as strategic planning of coastal defences, and to undertake the two activities in isolation, or with minimum contact, is to withhold from the spirit of holism to which many parties adhere. It is also an important point which is made in the Institution of Civil Engineers' own response to the Committee of Inquiry, although they suggest the best platform for co-ordination is the existing coastal cell groups.

Local authority initiatives

I wish now to briefly look at some of the responses that are being made by local planning authorities. My own group intends to publish a best practice guide in conjunction with a National Conference to explore in detail the findings from a National Survey of practice which has been completed by Posford Duvivier as consultant on behalf of Devon County Council in connection with their Exe study. The Exe study is a good example of the comprehensive approach now being adopted to explore linkages between land-use and marine activity, and to consider the 'marine interface' in a holistic way.

Another comprehensive approach worth highlighting is the example of the New Forest District Council's Coastal Management Plan, which is envisaged as a document subject to frequent updating - a learning tool between all parties - and therefore a means of guiding the adoption of policies in the statutory local plan. The approach covers everything from conservation, recreation and development through to coast protection and pollution and is intended to set out practical programmes of action. Very importantly, the plan proposes to identity all coast protection and flood defence works proposed or in progress and will ensure that environmental, planning and recreational issues are fully integrated into engineering programmes. Not surprisingly the plan refers to the Hampshire context which has been so ably set out in the Coastal Strategy for Hampshire and which has done so much to alert Government to the range of issues involved. Indeed, a paper on the Hampshire approach was presented at the 1989 conference by Peter Bell.

SETTING THE SCENE

Another county which has also undertaken a special County Study is Northumberland, but it would be wrong to overlook the existing policies contained in the Structure Plans of many maritime county planning authorities which have done so much to protect Heritage Coasts, Areas of Outstanding National Security, and other attractive areas over so many years from the landward side. In their reviews of these plans, greater consideration is now being given to estuaries and the probability of extending Heritage Coasts, although it must be recognised that the fact of such designation does not in any way render them invulnerable to pressures of one kind or another, as the National Trust knows full well in revitalising its Enterprise Neptune campaign to acquire more coastal properties and thus render them inalienable.

At the regional scale, I am impressed with the way in Wales the Strategic Guidance process has brought so many parties together despite my concern about the parallel cellular co-ordination of coast protection. Nevertheless their January 1992 Consultation Document has highlighted the key issues confronting coast decision-makers in Wales. The Welsh coastline extends for some 730 miles, of which 40% is Heritage Coast, although only 50% of coastal SSSIs are covered by the designation. In any review of boundaries there is scope for the inclusion of soft sediment coasts, including estuaries. The Heritage Coast Forum too has called for Management Plans based on three principles: temporal and spatial zoning of activities, the fixing of quantified management targets, and an annual work programme.

West Glamorgan experience

Much could be said concerning initiatives elsewhere - such as the 12 years of Sefton's Coast Management Scheme and the incorporation of appropriate policies in their Unitary Development Plan, the work of the Suffolk Heritage Coast and the Anglian Study. As a slice of typical reality I would refer to my experience as West Glamorgan's County Planning Officer. As a county of contrasts, with an industrialised seaboard just across the Swansea Bay from Gower AONB, with estuaries under pressure (or threat), with a barrage being constructed across the River Tawe, with an SSSI designed to protect sea birds only a few minutes away from the city centre of Swansea, it is an intriguing example. Both the Lough or Estuary, upstream from the North Gower ANOB SSSI, and the Neath Estuary, have been subject to barrage proposals despite Structure Plan policies, but both ran out of steam - basically for financial reasons - before a full environmental debate was necessary. An adopted Local Plan for Swansea, including Gower, has protective policies for the AONB and its Heritage Coast, together with a weighty Management Plan prepared in collaborative fashion with a number of parties.

But offshore, and under the sea, lie the Helwick Banks and reserves, not far from the Gower Coast. The consultation has been a classic example of the confusion that exists over final responsibilities for dredging licences. There has been stiff opposition from local councils, including the County Council as mineral planning authority. Considerable light on the dynamics of the situation, and the positive impact on Gower beaches, has come from the work of the Coastline Response Study from Worms Head (Gower) - Penarth Head done by Bryan Madge and Associates on behalf of the five district councils and British Steel. The study has also emphasised the vulnerability of the sand dune system at Jersey Marine to rising sea levels that could result from high rainfall, storminess and high tides.

Conclusion

In setting the scene, I trust I have shown that local planning authorities are alert to the complexity of issues within the coastal zone. In many respects the challenge is not so different from the co-ordination of land use and transport, which now ranges from the scale of global restraint to local traffic calming which might be seen as a form of soft engineering. Yet on the coast there are so many actors involved, and so many professionals on fast-learning curves. And that means the problem of the technical languages of so many specialists. I recently heard a distinguished marine biologist say quite bluntly, "The coast is up for grabs", referring, I believe, to the acquisitive interest of so many professional disciplines in making sure their own contribution is not overlooked, and that they have a major share of the action! And yet, more favourably, I also heard the Coastal Manager of a national geological institution say that never had he experienced so many professionals so keen to break down the barriers between their specialisms.

If this is the case, and I believe it is, then I believe it is a striking example of the new paradigm that is sweeping academic and professional life in the challenge and reality of environmental pressures, whether global, European, or national, and whether predictable or not. The truth is we must share our information, and more importantly our collective understanding.

Discussion

Universal agreement was expressed as to the need for changes in the management of the coastline with increased liaison and consultation.

The importance of comprehensive data acquisition was emphasised involving, in the case of the NRA, Anglian Region, an investment of £4 million, and the problem of financing such a scheme was stressed. The need for coastal authorities to undertake a monitoring programme was being recognised and methods of financing were under consideration. One speaker strongly recommended the involvment of a geomorphologist in the analysis of coastal processes.

If economic and environmental objectives were to be set there had to be a move to introduce environmental accounting. This was largely a political process. The difficulties of making an economic assessment of conservation benefits were stressed, recent and current work on the subject being mentioned.

The current grouping of coastal authorities in Wales was described, mainly based on sediment cells, together with the links to the Welsh Office/MAFF Forum. An essential component of Coastal Zone Management (CZM) was considered to be taking part in wider plans while promoting necessary works with least possible adverse effect.

It was pointed out that the draft Planning Policy Guidance Note on Coastal Planning (PPG) did not refer to marine archaeology. This aspect was not to be overlooked.

4. Natural change

DR J. PETHICK, Institute of Estuarine and Coastal Studies, University of Hull

Introduction
The problems associated with coastal change are as varied as the temporal and spatial scales over which they range. A coastal landform may be as small as a single cusp on a beach whose temporal adjustment to wave conditions may occur in minutes, or it may be as large as an estuary whose temporal adjustments to post-glacial sea level rise has proceeded over the past 10,000 years and may continue for at least that period into the future. Because of this enormous range, the identification of coastal change can, on the one hand, be an extremely difficult process, involving long term monitoring of subtle variations in, for example, mudflat morphology - or, it may involve a simple visual observation of the dramatic and almost instantaneous response of a beach to a storm. Yet understanding the reasons for such changes demands a philosophical approach to natural systems which appears to take the pragmatic coastal scientist or manager far from the important matters of process measurement or coastal defence works. Nevertheless, it can be argued, a failure to understand the reasons for coastal change has characterised coastal management in the past and has led to defence systems which are increasingly out of phase with the coastal environment, leading to costly maintenance or catastrophic failure. Increasingly, as part of our coastal management programmes, we are attempting to emulate natural systems rather than resist them - the so-called soft engineering approach, and this requires a degree of understanding which we have so far failed to obtain.

The ultimate reason for the observed changes in coastal morphology is simple enough - they are all responses to the necessity for coasts to absorb wave and tidal energy: coastal inputs which themselves are continually varying. The apparently simple dictum that 'coasts stop waves' contains within it a complex instruction for the development of coastal landforms over space and time. The recognition that coastal landforms absorb wave energy is perhaps not difficult, it is more difficult to appreciate that, in order to do so, they tend towards stable forms in which wave energy is absorbed without long term changes in morphology. The teleological nature of this type of statement requires some explanation. Why should coasts - or any landform - 'tend towards' morphological stability? And, if such an apparently goal-seeking hypothesis is accepted, why is it only too apparent to coastal engineers and managers that morphological change characterises every coastal form?

This paper examines the nature of both change and stability in coastal landforms as a response to wave energy. In particular it considers the conflict which exists between human use of the coastal zone, which demands and even assumes morphological stability and natural changes which threaten such use. Such conflicts appear to have been resolved in the recent developments towards coastal soft engineering in which coastal processes are emulated rather than resisted, as was the case in the older, hard, engineering approach. It is argued here, however, that such a soft engineering approach demands a much more detailed knowledge of the temporal adjustments of coasts to their energy environment than we appear to have acquired so far.

A model for natural change

The coastal zone acts as a buffer zone for the wave energy - both wind and tidal - which is generated over large areas of ocean. The passive acceptance of this energy leads to erosion which sets up morphological changes in the coast. Such changes will continue to occur unless, by random chance, a morphology is generated which absorbs the imposed energy without change. Such stable coastal forms, which can accept wave and tidal energy without long term changes in morphology, will eventually tend to predominate over the unstable forms so that a long term evolution towards stability can be predicted without any teleological overtones. Use of the term 'evolution' here is not intended as a synonym for 'development' but rather in the true Darwinian sense of the word. The coast does evolve a form which is adapted to its environment in precisely the same way as envisaged in the Darwinian evolutionary model. The environment for a given coastal landform consists of the energy inputs into the coastal zone and the inorganic and biological materials from which it is formed. Just as with a biological species, a coastal form which absorbs energy most efficiently, given the available materials, tends to survive at the expense of other less efficient forms. Again, just as with biological species, coastal forms interact with each other sometimes competing for available sediments and sometimes sharing materials in a form of symbiosis.

Although the analogy between organic and inorganic evolution cannot be taken too far, it is interesting, and perhaps instructive, to consider the parallels between the two in the temporal adjustments which they both make to environmental change. Since the energy and materials which make up the coastal environment are continually changing, the tendency towards stability by coastal forms must be seen as a dynamic stability in which adaptation to the environment is continually occurring. Biological species are characterised by three main adaptations to such environmental change. First, a species may be relatively insensitive to short term changes such as diurnal temperature changes or short term meteorological variations so that no corresponding variation is set up in the size or location of the species; such insensitivity represents one scale of adaptation of the species to the environmental conditions. Second, the species may respond to medium term events, such as seasonal or episodic climatic changes, by adjustment in the numbers of organisms within the species or by locational changes such as migration. Once the limiting factor is past, the species can return to its optimum size or location so that no long term changes are necessitated. Last, the species may respond to long term changes in its environment by evolving a different form altogether. These three temporal adjustments are nested one within another so that long term evolutionary change can take place while short and medium term responses continually adjust to it. Each of the three adjustments also sets up a response in neighbouring species, a response which is affected by competition for space or food.

Natural change in coastal landforms

Although it may be argued that there can be no parallels between an organic species and an inorganic landform, yet in neither case is there any question of motivation or teleology. The processes of evolutionary adaptation, whether of an animal or a coastal form, takes place by the positive process of random mutation which in the coastal case may be the chance generation of a distinct morphology - a steeper or flatter beach slope for example. These random morphologies may then be selected - that is they will persist - if they happen to fit the environmental conditions. If not they will be eradicated by death in the case of an animal, by erosion in the case of a coast. In each case the environment itself cannot be expected to remain steady so that this process of adjustment is a continuous one.

A coastal landform, therefore, responds in precisely the same way to its imposed wave energy and sediments as a species reacting to imposed conditions of food, space and temperature. The end point of change in both cases is long term stability, which is not derived through any teleological process but merely represents the simple probabilistic statement that stable systems endure while unstable systems are transitory. As in the biological case outlined above, three scales are proposed for such changes

(a) a short term, resistant phase in which the coast is not sensitive to small scale changes in the wave or tidal energy variations

(b) a medium term, cyclic phase in which the coast responds to periodic energy variations

(c) a long time period over which secular changes occur which are superimposed on the previous scales.

PRESSURES ON THE ZONE

It appears that the resistant phase of such a classification is one which is most easily appreciated by us. Indeed, we expect coasts to be resistant to wave attack and are surprised, even annoyed, when they depart from such expectation. If coastal forms are observed to be in the process of changing, however, we automatically assume that such changes are secular - that is progressive. Thus the most difficult part of the model for us to accept is the periodic phase of coastal adjustment. It is at this scale of coastal change where both layman and scientist alike experience problems of interpretation and it is, therefore, one which is central to the present analysis.

Resistant or insensitive forms

Coastal forms may be resistant to energy conditions for one of two reasons. First, the coastline may be composed of materials which are inherently stronger than the applied stress by waves or tides. Coastlines such as the granite cliffs of the Land's End Peninsula or the fjord coastline of Norway may be regarded as 'fossil' features in that they owe their morphology to some previous energy environment rather than to any adjustment to their immediate environment. Alternatively, the resistance of a coast to energy variations may have evolved as changes, initially set up by an imbalance between stress and strength, present a sequence of different forms, the most resistant of which will be selected by the environmental conditions. Thus erosion of a cliff may result in a deposit of unconsolidated sediments at its toe, whose constituent particles absorb low wave energy levels by internal friction. The resulting beach profile will be insensitive to short term variations in the received wave energy, a profile that resists, rather than adjusts to, the imposed wave energy so that no gross morphological change takes place within this time frame. Similarly, salt marshes develop a marked resistance to wave energy once they are colonised by vegetation. The increased roughness length provided by the vegetation prevents high frequency, low magnitude waves from eroding the marsh surface so that both biological and inorganic forms achieve short term stability.

Periodic change

Periodic or cyclical change, as discussed above, is the most difficult aspect of coastal adjustment to measure or to understand. It is important here to distinguish between episodic progressive changes, brought about by storm events for example, and cyclical change in which the coastal form responds to a storm by altering its morphology but then recovers and the morphology returns to its previous state. Thus, cliff erosion may proceed episodically but progressively, so that no recovery or retrogressive change is present. Fig. 1 shows the temporal sequence in retreat of a cliff on the Holderness coast.

A three year periodicity is produced by the interaction of episodic toe erosion by waves triggering mass failures which subsequently protect the toe from further erosion for a limited period. This type of intermittent progressive change is quite different from the cyclic change of some coastal landforms which involve a recovery, partial or complete, from each successive erosive phase. Beaches, for example, develop dissipative profiles, which are longer and flatter, during high energy wave events but, after the forcing event is over, gradually return to the steeper reflective profile. Fig. 2 shows a temporal sequence of beach profiles located at the toe of the cliff shown in Fig. 1.

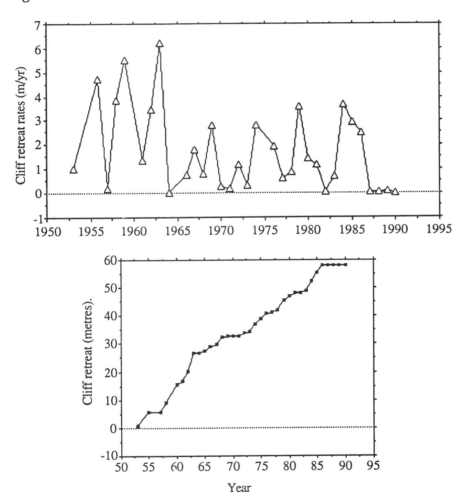

Fig. 1. Retreat of the Dimlington Cliff, Holderness 1953-1990. A three year periodicity in retreat rates (upper figure) results in a stepped spatial retreat (lower figure)

PRESSURES ON THE ZONE

This cyclicity, in which forms which have responded to high energy events return to a low energy response once the extreme event has passed, presents us with a major problem of monitoring, since measurement periodicity must be more rapid than the, initially unknown, morphological frequency. The problem of interpretation is even more complex, since it is difficult at first sight to see why a form which is resistant to a high stress should subsequently return to a form which is resistant only to low applied stress.

The behaviour of organic populations provides the answer. Periodic reductions in the numbers of individual organisms in a species may be forced by episodic environmental deterioration, the reduced numbers may be better

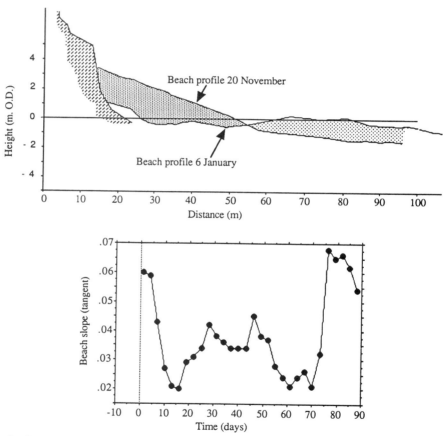

Fig. 2. Variations in beach profile at Dimlington, Holderness. Upper figure shows typical profile response to a storm event. Lower figure indicates cyclicity of response during a three month winter period

able to withstand the lower availability of food during the period of privation, but once better conditions return the species recovers and returns to its optimum numbers. The periodicity is, of course, an expression of the balance between supply and demand for food. At the coast this process is brought about by the balance between erosional and depositional processes. Sediments accumulate at the coast during periods of low wave or tidal energy and form a morphology which expresses the balance between material strength and environmental stress. A mudflat, for example, has a surface which lies at the critical threshold of erosion. If more sediments accumulated on this surface, then wave shear stress would increase because of the shallower water depths and erosion would re-establish the previous form. If the surface is lowered, then accretion will commence due to the lower wave stresses produced and the initial surface will again be restored. Fig. 3 shows this process on a salt marsh surface; here the vegetation cover increases the resistance of the marsh surface to wave shear by increasing the roughness length of the velocity profile, so that vertical erosion takes place only during extreme wave events.

The probability of occurrence of a high spring tide which will cover the marsh, combined with a 1 year return interval wave, gives a 1 in 30 year probability for the marsh shown here. Recovery of the marsh occurs once the storm event has passed, since the lower surface experiences more tidal inundations and therefore higher accretion rates than it did before the vertical erosion took place. The result is that this marsh is able to recover from the storm event within 5 years. For coastal landforms generally, the response to a higher wave energy event is to increase the surface area over

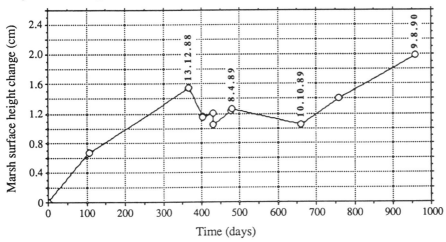

Fig. 3. Erosional response of salt marsh surface to storm event of 20.12.88 and its subsequent depositional recovery. Dengie Peninsula, Essex.

which energy is dissipated, a change effected by the redistribution of sediments involving both erosional and depositional processes. Once the high energy forcing event is past, however, the morphology is 'over-efficient' for the subsequent low energy waves and creates conditions in which accretionary processes act to restore the previous low energy form.

The recovery of coastal forms from high energy events depends, therefore, on the availibility of a sediment supply and a depositional process which can affect a return to a low energy morphological response. Depositional processes are stimulated by the erosion of coastal forms - since lower surfaces experience deeper water depths and consequently lower bed shear - nevertheless this seems to imply that sedimentary processes are in some way programmed to respond to the need for landform recovery. That this cannot be so is self-evident, but a corollary suggests that the identification of sediment transport processes cannot be used to predict landform development. Thus, coastal suspended sediment transport involves a large pool of material, only a small percentage of which may be involved in deposition so that there is no causal connection between the transport process and coastal development. This means that measurement of process may provide only a necessary, but not sufficient condition to explain coastal form, while the alternative method of measuring sediment budgets also fails due to the attempt to isolate the extremely small volumes involved in landform deposition from the very large volumes carried in the sediment pool.

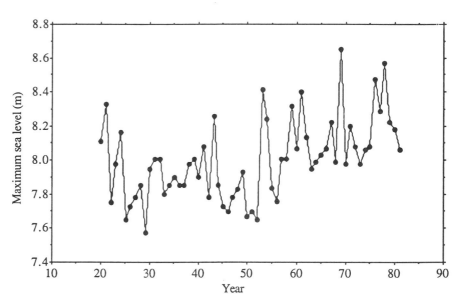

Fig. 4. Maximum annual sea level at Immingham, Humber Estuary 1920-82

The behaviour of non-sedimentary coastal forms is of course quite distinct from that of the sedimentary forms described above. Cliffs, for example, although they experience an periodic erosional response to variations in wave energy, do not recover from such wave events, since an accretionary phase is missing. Thus cliffs either are resistant forms, the result of some fossil process, or they show a long term retreat marked by periodic erosional phases linked to wave energy inputs. In some cases, where erosion produces a large debris input to the cliff foot, this periodicity may be emphasised by the debris mantle at the foot providing protection from subsequent storm events as shown in Fig. 2. Yet true cyclicity is clearly impossible in a cliff system where the sediment transport passes across a process divide - that is from sub-aerial to littoral - so that landform recovery is prevented. In some cases, notably micro-cliffs in inter-tidal deposits such as mudflats and salt marshes, cliff recovery from storm erosion is possible since sediment transport is able to move both up and down the cliff face. Such conditions are rarely met, however, and even in tidal salt marshes, the presence of a cliff normally indicates either a static resistance to wave energy or a progressive long term retreat without any intervening cyclicity.

Secular or geologic change

Longer term changes in the imposed coastal environment, such as the post-glacial change in sea level, result in the gradual evolution of a variety of different coastal forms which themselves are able to react to short and medium term energy variations. As noted above, secular changes may be manifested as a slow progressive change superimposed on the cyclic response to environmental variations or they may be periodic, in response to a similar periodicity in the environmental forcing. Fig. 4, for example, shows the strong periodicity present in storm surges within the Humber Estuary. Such surges result in periodic erosional phases along the estuary shore which are progressive but not cyclic since they are not associated with any intervening recovery phase but merely with quiescence.

Most secular changes at the coast are due to past changes in the energy of the coastal environment, principally those associated with the post-glacial rise in sea level, but there is some speculation that a long term change in coastal materials may also be involved. Thus the accumulation of large amounts of no-cohesive sediments on the coastlines of the world may be a short term response to the post-glacial rising sea level and this material is gradually being moved offshore back to the shelf. This may mean that beaches are an ephemeral feature of the coastline and will eventually be replaced by rock abrasion platforms which represent a much more stable morphological adaptation to the dissipation of wave energy. If such an hypothesis is accepted, it is interesting to note that it involves a massive increase in the resistant phase of coastal change and a concomitant decrease in the rate of periodic change. It is interesting to speculate further that such a development would mean the increase in the resistant phase of coastal response to energy inputs, at the expense of the periodic phase, suggesting that periodic change is itself merely an ephemeral stage in the adjustment of a coastline to its environment.

The recognition of long term progressive change is one of the major problems facing the coastal scientist. Cyclic changes may have periodicities measured in hundreds of years, especially in low energy environments or large scale landforms such as estuaries, so that, given the short time period over which we have records, these become impossible to distinguish from any long term secular signals which may be present. Fig. 4 shows a relatively straightforward case of a periodic signal, with what appears to be a long term secular change superimposed on it, but even this relatively long term (60 year) data record is not extensive enough to preclude the chances of confusing secular and periodic change. Indeed any such attempt at distinguishing between the two may be dismissed as meaningless, for all change becomes periodic if a sufficiently long time scale is defined. Nevertheless, a pragmatic approach to coastal management demands that not only is the presence of long term progressive changes recognised, but also that the rates of change are quantified so that appropriate action may be taken to protect the interests of coastal users. One of the most important aspects of such an approach concerns the predicted changes which may result as a response to sea level rise over the next century - a form of long term, secular coastal change which is discussed below.

Sea level rise

The sea level rise predicted to affect world coastlines as a result of global warming will set up secular changes in coastal morphology which may well be extremely damaging to our present use of the coastal zone. Although sea level rise cannot proceed indefinitely, the predictive models so far available to us have attempted only to estimate the changes expected by some arbitrary date - usually the year 2050. The problem facing coastal geomorphologists is whether it is possible to predict the morphology of the coastline while the forcing event is still proceeding.

The effect of sea level rise on coastal forms may be sub-divided into three elements

(a) an increase in the periodicity of extreme water levels - since the probability of storm events affecting the upper shore is the combined probabilities of waves and water levels, this in turn means that an increase in the periodicity of extreme wave events at the shore will result

(b) an increase in wave energy at the shore brought about by the increase in water depths and a consequent decrease in wave attenuation

(c) a change in the boundary conditions at the coast - this includes the obvious effect of landward migration of the extreme water level forming longer tidal channels, increasing the size of bays and so on; it also involves a change in the position of the deep water/shallow water demarcation and consequently in wave refraction patterns.

The response of the coast will be for a long term adjustment to increased energy, brought about by progressive shifts in the periodic morphology. The increase in water level predicted for the year 2050 is presently put at <1 m; most coastlines of the world regularly experience such water level increases due to surges, fresh water inputs or seasonal temperature increases, so that the predicted long term sea level rise will not provide new environmental conditions but merely an increase in the frequency of present events. An increase in the frequency of extreme events means that the time available for the coastal system to recover between each event is reduced so that progressive change is superimposed. Prediction of the morphological response to such an increase will depend on our ability to recognise and quantify the present rates of such periodic changes. The coastline will, given sufficient time, adjust to any changes in sea level forced upon it; indeed an increase in water depth usually leads to an increase in deposition rates and a decrease in the periodicity of the coastal response so that continuous morphological adjustments to rising sea level are maintained. If the rate of sea level rise is so rapid compared with the periodicity of coastal response that recovery from individual storm events is not possible then progressive deterioration of the coast will set in. It is therefore the rate of sea level rise, rather than the absolute increase, that is crucial to the prediction of coastal response and that requires our urgent attention.

Coastal symbiosis

The interaction between coastal landforms involves one of the most complex issues in coastal change. The temporal nesting of resistant phases, and periodic and secular changes for a given coastal form have already been explored in this paper but here attention is turned to the problem of interactive change in spatially separated coastal forms. Such interaction takes place where energy and sediment moves from one coastal area to another. Normally the energy flux between adjacent forms is unidirectional although in many tidal systems an interchange of energy can take place between adjacent parts of the tidal system. Morphological response, however, depends upon sediment movement, and here the differences between unidirectional sediment flux and sediment interchange is crucial. Where only unidirectional flux occurs, as in the cliff debris movement described above, the resultant morphological response is to secular change. Where sediment exchanges

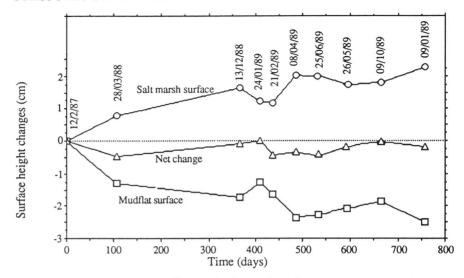

Fig. 5. Vertical changes in mudflat and salt marsh surfaces, Dengie Peninsula, Essex. The marked inverse association between the two morphologies is especially noticeable during the easterly storm event of Dec/Jan 1988/89.

take place, however, a cyclic response may be produced linking two coastal forms.

The beach-sand dune system may be taken as an example of a complex interaction between adjacent coastal forms having quite distinct reaction times to high energy events. During low wave energy events, sand is deposited on the upper beach and provides a source for aeolian sand transport which accumulates in the supra tidal zone as sand dunes. During high energy events, however, the beach response is to become flatter and to lengthen, so that the seaward dunes are eroded. This provides an additional source of sand which is transported seaward and deposited on the lower beach as well as extending the beach landward thus increasing the length of the absorbent surface. The dunes therefore act as a form of long term insurance for the beach which is cashed in during storm events.

A similar system is to be observed in the mudflat-salt marsh interaction. Here the salt marsh is able to act as a sediment store during low energy events due to the protection from wave energy afforded by the fronting mudflat and to the presence of a vegetated cover. During extreme storms, however, the normal mudflat morphology is inadequate to absorb the high wave energy and its upper profile erodes by cutting into the salt marsh. In severe storm, vertical erosion of the salt marsh surface may also take place. As in the case of the sand dune-beach system, this erosion supplies sediment which is deposited on the fronting mudflat during the storm thus increasing its energy absorption properties. Once the storm event is over both mudflat and salt marsh recover as accretion returns the morphology to its previous, average state. Fig. 5 illustrates this type of inter-dependency: the effect of storm erosion on a cohesive inter-tidal profile on the Essex coast is shown to result in a temporal sequence of erosion and accretion on the mudflat which is matched by an almost perfect mirror image in the response of the salt marsh.

The inter-dependence of coastal systems such as the marsh-mudflat shown in Fig. 5, means that each separate part is more stable when it is associated with the other. These dual systems are an aspect of coastal morphology which is often overlooked in our management approaches: salt marshes depend upon the energy-absorbing properties of the fronting mudflat during low energy events while, in turn, the mudflat relies on the supply of sediment and the spatial extension provided by marsh erosion during storms. Thus the combined beach-dune or mudflat-marsh systems stand a better chance of survival than they would as separate landforms - a form of symbiosis which is as complex and as stable as many biological relationships.

Human use of the changing coast

Although it is not the role of this paper to consider the impact of human activity on the coast, nevertheless it may be apposite here to consider briefly the way in which we have tended to respond to the complex changes in time and space considered above. Human use of the coast in almost every case requires morphological stability. We tend to regard any form of change as inimical to our interests and to characterise processes such as erosion as a type of crime committed against us by natural, but malign, forces. Moreover, due to the difficulties of deciphering one signal from another, we have failed to distinguish between cyclical and secular change except for very short term periodic changes - such as those exhibited by high energy coasts such as sand and shingle beaches. Instead, we have classified coasts as those resistant to change and those exhibiting secular change - the hard and soft coasts of many early 20th century text books. We have also failed to recognise that all change leads to stability and that, consequently natural processes are working towards the same ultimate end as we are. Ignoring this, we have characterised all change as bad and have developed two types of response. First, in most cases we have chosen to ignore the risks involved in locating on a changing coastline. Second, where possible we have attempted to increase the resistance of coastlines by artificial means. We have not devised mechanisms for adjusting to, or even avoiding, coastal hazards and until very recently we have not attempted to accelerate the natural development towards coastal stability.

PRESSURES ON THE ZONE

Avoiding or adjusting to areas of coastal change is more difficult than it seems. There are many pressures and incentives on humans to locate on the coast regardless of the risks involved; they include defensible positions, communications, fishing, flat land, good soils and, more recently, recreation. Thus even in areas such as Bangladesh where the estuary bank erosion rate may be as high as 100 m per year and flooding is a major danger, human settlement takes place, while the number of villages that have disappeared on the eroding east coast of Britain in historic times indicates that a coastal location often takes precedence over long term stability.

Increasing the resistance of a coastline has been our main response to natural change. The results on high energy coastlines have generally been disastrous. The impediment to sediment movement produced by such hard defences and the effect of this on adjacent coastal systems are now widely appreciated. The tendency for hard defences to develop potential instability in the protected coast is also well known. An increase in resistance of the coast to wave energy also means that short to medium term periodic adjustments to storm events are impeded. Since long term secular change is brought about by the gradual and progressive change in the periodic adjustment of the coast this means that the protected coast becomes increasingly out of phase with the imposed energy regime leading to a demand for resources to increase the level of the defence or its catastrophic failure.

In many cases, of course, the protection afforded by hard engineering is essential despite the increasing cost of maintenance in the face of long term increases in the potential instability of the coast. In such cases, the effect of an impeded sediment supply to adjacent coastal systems may add to the economic and social costs involved and should be taken into consideration. One of the problems of assessing such costs is the difficulty of timing the response of the system to change.

Finally, our response to coastal change has been impeded by our failure to appreciate that cyclical change is a fundamental response of the natural system to its environment. This is less true of high energy systems such as sand or shingle beaches than it is of low energy mudflats or estuaries. High energy beaches respond rapidly to changes in the energy environment and it is obvious to even a casual observer that beach slopes are lowered within a few hours during storms but steepen again over a period of weeks subsequent to the storm. In low energy environments changes may take much longer. The lowering of salt marsh surfaces by wave action may occur at intervals of 10 years and the recovery from such events may take 5 years. Recognition of periodic adjustments of this type is difficult without long term monitoring and consequently the changes involved are often mistaken for progressive changes.

Our failure to appreciate medium term cyclicity in coastal change is also apparent in our recent attempts to conserve coastal habitats. Coastal habitats such as sand dunes or coastal wetlands such as salt marshes and mangroves are regarded as major environmental resources and as such are protected by international treaty as well as by national law. For example the 1979 European Berne Convention on the Conservation of European Wildlife and Habitat and requires contracting parties to: 'take appropriate and necessary legislative and administrative measures to ensure the conservation of habitats ... and the conservation of endangered natural habitats'. Of particular relevance is the obligation on parties: 'in their planning and development policy' to 'have regard to the conservation requirements of the areas protected ... so as to avoid or minimise as far as possible any deterioration of such areas.'

The fact that these requirements were introduced to conserve species of plants and animals ignores the fact that their habitats are ephemeral components of the coast. Any attempt to preserve these habitats will, in the long term, produce the opposite effect of forcing disequilibrium on the coastline by preventing its evolution towards stability. Attempting to prevent natural periodic changes in these ephemeral wetlands could be likened to an attempt to prevent bird migration - the change is necessary for survival. Nevertheless, since so much natural habitat has already been removed, that which remains becomes of great significance. We should, therefore, consider the possibilities of re-locating or re-generating such valuable habitats so that the effects of any natural losses which do occur are ameliorated.

Conclusions

The recognition that natural changes are essential for coastal adjustment to a constantly varying environment is one which we have been slow to appreciate. In particular, we have failed to understand the complex periodic changes which characterise many coastal forms, since such changes include both episodic progressive change, cyclic change and the superimposition of both. Instead, we have tended to classify coasts as either resistant or eroding and have reacted by attempting to increase coastal resistance - resulting in coastlines which are increasingly out of phase with their environment and which require costly maintenance programmes. If we are to adopt a more sensitive approach to coastal management in which we attempt to emulate, rather than resist the natural systems, then we must rapidly acquire an understanding of the periodic adjustments of coasts to energy inputs and allow, in our management programmes, both sufficient space and time for these to operate.

5. Coastal heritage

M. KIRBY, Countryside Commission

This paper concentrates on four initiatives with which the Countryside Commission is closely associated and which help sustain coastal heritage. Two of the initiatives are a strategic first: the definition of Heritage Coasts under Circular 12/72 'The Planning of the Undeveloped Coast' and the promotion of a European framework for coastal zone planning and management as recommended by the European Workshop on Coastal Zone Management which the Commission and the National Trust organised in April 1991, with support from the European Commission, the Department of the Environment and British Petroleum.

These strategic initiatives complement two major programmes of practical action: the Countryside Stewardship programme, which has as one of its main elements a programme for coastal land, and the Commission's moral and financial support, over the last 30 years for the acquisition of coast by the National Trust via Enterprise Neptune. The Commission also funds the management of this land by way of annual grants to the National Trust.

To appreciate the different aspects of coastal heritage, a definition of the coastal zone is necessary. The definition I offer was developed at the European Workshop on Coastal Zone Management and endorsed by the Countryside Commission in June 1991. The definition (and the recommendations of the Workshop) were presented to the European Commission's European Coastal Conservation Conference hosted by the Dutch Government in November 1991. The conference accepted it as a useful working definition; it is flexible and has considerable utility in the very wide range of interests and pressures that affect the coastal zone. The definition states

'The coastal zone is a dynamic human and natural system which extends to seawards and landwards of the coastline. Its limits are determined by the geographical extent of the natural processes and human activities that take place there. Coastal zone management should extend as far inland and seaward as is required by the management objectives.'

In developing a complete list of all the heritage qualities in the coastal zone one could usefully look at that definition, in particular the reference to 'natural processes and human activities...' We are familiar with fine landscapes and important wildlife habitats being described as heritage. However, one way of securing their effective conservation is to understand

PRESSURES ON THE ZONE

better and manage more effectively those heritage qualities which are more directly a consequence of human activities.

The coastal zone is important in social, economic, political and military terms. It has a long history of settlement, trade, commerce, industry and exploitation of its natural resources, both marine and terrestrial. We still refer to ourselves as "a seafaring nation" or as "an island race" which, although not fully accurate, indicates the cultural importance of the sea and the coastal zone to how we view ourselves historically and politically.

Much of our literature, music and painting signals a strong bond between people, and the fact that we live on an island. The coast and the sea are important to our political and economic life and there is a strong emotional bond between people and the resource. We respond to the images conjured up by John of Gaunt.

> 'this scepter'd isle'
> 'This fortress built by Nature for herself Against infection and the hand of war,'
> 'This precious stone set in the silver sea' [Richard II Act II scene 1]

There is a strong romantic and historical perspective of coast and sea in our national heritage.

Pressure for change makes it difficult to conserve or sustain all of these heritage qualities, all of the time. However, it is an essential first step towards integrating coastal zone planning and management, to appreciate and document all of the heritage qualities within the zone before identifying those which are capable of being influenced by planning and management. Through a better understanding of the range and importance of these heritage qualities in the coastal zone, we can arrive at more relevant and therefore more effective coastal zone plans.

Take, for example, fishing in inshore waters. Over-exploitation of the resource and the reduction in the resource by pollution jeopardizes fishing as an economic activity. But fishing, fishermen and fishing boats are part of our cultural heritage. Visiting one of our many small harbours 30 years ago one would see a few holiday makers watching the fishing boats coming and going, and landing their catches, attracting little knots of people around the boxes of fish, crab and lobster. Today, all too frequently the picture has changed. A surge of holiday makers armed with cameras and cam recorders jostle around an ever diminishing but already small number of local fishermen, hanging on in an industry where the financial margins grow thinner every day. The tourism industry frequently overwhelms and/or replaces primary industries such as fishing with a consequent impact on housing for locals, local schools and other local services.

Harbours without fishing boats and fishermen - replaced by yachts and yachtsmen - represent a change which has an impact on our coastal heritage.

These changes might be both beneficial and inevitable. The point I make is that we have to be aware of these changes and to evaluate them objectively. The loss of or damage to the coastal heritage must not happen by default. Integrating planning and management requires a synoptic appreciation of all relevant heritage qualities, including our cultural heritage.

In this respect it is interesting to note that at the European Workshop one of the three key areas for action that should be addressed by the coastal zone plan was the 'human communities that depend on a healthy productive coastal and marine environment'.

Since 1972, in partnership with local authorities, 44 stretches of coastline have been defined by the Countryside Commission as Heritage Coasts: some 1,496 kilometres. These represent not only the finest coastal landscapes but also contain about half of all the coastal SSSIs, and their popularity for recreation and access is very significant to the tourist industry. In most cases special arrangements have been put in hand for their planning and management. One third of all heritage coastlines is owned by the National Trust, a fact which I will return to later in this paper.

The purpose of defining Heritage Coasts is to focus attention on their management. In most cases the Development Plans covering Heritage Coasts afford them an acceptable degree of protection from development. But a combination of factors, particularly changes in agriculture and the rapidly increasing pressure from tourism, requires their effective management. That is management in the sense of estate management: action on the ground, often small-scale to conserve landscape and wildlife and at the same time to enable people, in ever increasing numbers, to enjoy the cost without damaging the resource.

Conserving and enhancing the quality of Heritage Coasts and facilitating their enjoyment requires co-ordination and co-operation from a wide range of organisations with responsibilities along the coast. Notwithstanding that Heritage Coasts do not deal with the marine environment nor with the urban or urbanized coast where lie many of the most serious problems, the Heritage Coast management approach is effective. This approach, an approach based on co-ordination and co-operation, applied to the wider canvas of the whole coastal zone, is the right strategy to achieve significant progress in integrating planning and management within the coastal zone.

A key feature of the Heritage Coast approach is the identification of zones within which a different emphasis of activity and management can be proposed and resourced.

Given the wide range of conflicting interests which use and exploit the resources of the coastal zone, a zoning policy for development and management is useful. Some tidying up of the administrative complexities that currently constrain effective action at the moment would be welcome, but in the UK the likelihood is that change will be slow and evolutionary rather than radical and rapid. Thus, the Heritage Coast zoning approach could usefully be extended to the whole coastal zone.

The second initiative with strategic relevance is promoting the need for a European framework for planning and action. The European Workshop on Coastal Zone Management brought together delegates from the 11 Member countries of the European Community with coasts, together with Sweden. The Workshop is one link in a chain of European events developing the case for the integration of planning and management in the coastal zone throughout Europe. Simultaneously, and within a framework agreed by the European Community, each European coastal nation should prepare a comprehensive strategy for its coastal zone. Each national strategy would require action at the local, region and national levels. This action would embrace

(a) planning: that is providing strategic objectives based on a survey of all relevant environmental and socio-economic factors

(b) management: a means by which strategies should be implemented and monitored, including setting up the necessary administrative and regulatory structures.

The UK has the longest main land coastline of all EC members. Our development plan system stands comparison with the best. But strategically it is not only important that we make progress but it is also important to secure complementary high standards of planning and management throughout Europe's coastal areas. Many UK residents own wholly, or in part, property on or close to coastlines of other European states and millions holiday on these coasts or go sailing in the seas around them.

Now two initiatives of a more tactical nature. First, Countryside Stewardship, which is a new approach to conserving countryside, including coastal lands, and creating opportunities for people to enjoy it. Stewardship is a Government initiative launched as a pilot scheme on an experimental basis in June 1991. The Countryside Commission is responsible for designing and operating the experiment in collaboration with the DoE, MAFF, English Nature, and English Heritage. The budget for the first three years of the experiment (1991-94) is about £20 million.

Stewardship adopts the principles of the Environmentally Sensitive Area approach and then exploits their further potential in a number of ways

(a) eligibility for the scheme is universal to all who own or manage land, including conservation bodies and local authorities

(b) the scheme does not require boundaries and can be applied to the whole of the coastal zone

(c) the scheme is flexible - while agreements are to a standard form, the management guidelines permit local characteristics, practices and traditions to be followed, where appropriate and beneficial

(d) the scheme is discretionary: to landowners and managers it offers flexibility and choice, but to the Commission it allows selection of schemes with the best potential for environmental improvement, public enjoyment and value for public money.

Stewardship is a national scheme, initially targeted at five English landscapes of which coastal land is one. In the first year we received a strong surge of coastal applications covering in excess of 10,000 acres, of which 3,500 acres offer access. If added together, the coastal applications covered approximately 35 km of coast.

Countryside Stewardship for coastal land aims to

(a) sustain and restore existing areas of coastal vegetation and the wildlife it supports
(b) restore and protect characteristic landscape features
(c) create and improve opportunities for people to enjoy the landscape and its wildlife.

Within the coastal zone the Commission has targeted Stewardship towards existing areas of coastal vegetation, for example

(a) on cliff tops and salt marshes
(b) towards traditional grassland marsh where the re-introduction of management would be beneficial
(c) to cliff-top arable land or grass leys in areas of high scientific value
(d) to arable land or grass leys that link or extend fragmented remnants of natural or semi-natural vegetation
(e) to landscapes particularly rich in historical or archaeological remains
(f) to land that offers opportunities for people to enjoy landscape and wildlife through new public access.

In this first year we have targeted a combination of landscape, wildlife and historical heritage. This blend of objectives will continue over the next four years of the Stewardship pilot scheme. Unlike the other landscape targets where access is discretionary all coastal land proposals must include provisions for public access, either new access through Stewardship or through existing public rights of way. We do not insist that the access must extend over all the coastal land entered into the scheme, particularly where that land contains historic monuments or areas of importance for wildlife sensitive to human intrusion. But a key element in securing public benefits from coastal land in Stewardship is to ensure that the public can enjoy their heritage of landscape, wildlife and history. The link between enjoyment and awareness, and awareness and action is important. Securing coastal heritage in the longer term is achieved more effectively if the public use and enjoy the coast and are directly affected by the quality of its environment.

PRESSURES ON THE ZONE

Simply by raising the issue of pollution on beaches and awarding clean beaches a 'Blue Flag' the public have become more environmentally aware and exercise consumer choice. Whether it is Blackpool or Scarborough or a remote beach in Cornwall or Cumbria, the public now look for the 'Blue Flag'. The scale and priority of some water and waste management programmes have been modified to meet the standards of the 'Blue Flag'. The public's use and enjoyment of the coastal zone is a powerful force for the effective conservation and management of the coastal heritage.

The second initiative with which the Commission is associated is the acquisition of coast by the National Trust via Enterprise Neptune. The National Trust now own over 520 miles (840 kilometres) of our coastline, covering an area of land in excess of 115,000 acres (47,000 hectares), much of it subsequently declared by the National Trust as 'inalienable'. There is a close correlation between the Trust's acquisitions through Enterprise Neptune and Heritage Coasts. The Countryside Commission gives high priority to grant aid for the acquisition of Heritage Coasts by the National Trust. Other conservation bodies such as County Naturalists Trusts and the RSPB also acquire important stretches of coastline, sometimes with specific grant aid from the Commission. Protective ownership is a most effective way of securing the conservation and enjoyment of our coastal heritage.

Following acquisition, the National Trust prepared a management plan setting out how landscape, wildlife, history and visitors are to be managed. On an annual basis the Commission provides grants to the National Trust for conservation, access and recreation management on these properties. This might include a car park, information and interpretative leaflets, guided walks, maintaining footpaths, restoration of stone walls, woodland management, tree and hedgerow planting. The point is that pressures on the coastal heritage require positive management on a permanent basis.

Development planning in itself will neither sustain nor conserve the heritage qualities of the coastal zone. Only positive management, especially by those who own and manage the land, can safeguard in the long term the rich heritage of landscape, wildlife and history which represent but three of the many heritage qualities within our coastal zone.

6. Regenerating the developed coast

L. SHOSTAK, Director, Conran Roche Planning

Introduction
Most UK coastal cities and towns have significantly underutilised or derelict waterfront sites. Many local planning authorities and development agencies have already recognised the important role which these sites can play in comprehensive efforts to promote economic regeneration.

Whilst most of the recently published draft Planning Policy Guidance Note on Coastal Planning refers to the undeveloped coast, the draft does recognise the opportunities offered by the developed coast which

'... may provide opportunities for restructuring and regenerating existing urban areas, reclaiming derelict areas and accommodating new development, provided that due regard is paid to the risks of erosion or flooding. Where development requires a coastal location, the developed coast will usually provide the best option from an environmental perspective.'

In this paper, I draw on CR Planning's recent and current experience with efforts to regenerate the developed coast. This includes preparing

(a) development plans for two key areas in the Cardiff Bay urban development area; implementation of these large residential, office and industrial and leisure schemes is now underway

(b) a strategy for the regeneration of Southport's seafront to strengthen its role as a centre for tourism and leisure; this scheme has been the subject of considerable local controversy

(c) a master plan for the redevelopment of the North Dock and adjacent coast in Llanelli; the gap between infrastructure and environmental protection costs and forecast development values is currently too high to permit implementation to proceed

(d) a detailed master plan for Royal Quays in North Tyneside; implementation of this residential, retail, leisure and industrial development is now underway.

In three of these locations, our plans have had to take account of natural or planned changes in the relationships between sea and potential development sites. The construction of the Cardiff Barrage will lead to the creation of prime development sites surrounding the new Bay formed behind the Barrage (Paper 12). In Southport, the recommended development strategy focused on the lake between the beach and the edge of the built-up area as the sea has retreated from the former bathing beach (Paper 16). In Llanelli, a redundant dock will be the heart of a new residential community but the current tidal regime limits the development potential for marina facilities.

We are now preparing a regeneration strategy for the heart of Brighton's seafront. In addition, we have advised the authorities responsible for selecting Adelaide as the site for the joint Japanese-Australian project to attract technologically advanced industries to South Australia. The selected site is a degraded estuarine site with considerable ecological significance.

In this paper, I intend to draw out lessons from these experiences. In order to respect commercial confidentiality I shall describe this experience in general terms rather than with specific relevance to each project.

I will then suggest guidelines for the local authorities and professionals involved in efforts to promote the regeneration of the developed coast. I will pay particular attention to the opportunrty offered by the draft Planning Policy Guidance on Planning.

Lessons from experience
The developed coast offers considerable economic potential

In the UK, as in most countries, the coastal zone offers an environment for living, working, and recreation. Coastal sites in existing urban areas which were once used for industrial or port use offer massive potential for leisure and tourism, retail, residential and premier quality office uses. Many seaside towns are littered with underutilised or even derelict sites which were once at the heart of a thriving resort.

The developed coast - existing or potential waterfront sites in built-up urban areas - can play an exceptionally important role in any effort to regenerate an urban area. When serviced, and set in a positive planning regime, the developed coast offers considerable potential to attract new investment and enhance a community's economy. This investment and economic potential has three elements.

Tourism and conferences. Britain's seaside towns are still popular tourist destinations and conference centres. Yet, as the economic integration of Europe gains momentum, we must recognise that the competition from European cities will become more intense. Le Touquet, Schevingen, and Blankenberge and Ostende, and many others, will become even more popular alternatives. If Britain's seaside resorts are to compete effectively for a fair share, or more, of the tourism and conference markets, they will have to construct better hotels, conference facilities, and visitor attractions and invest heavily in the public realm. Key coastal urban sites are the prime investment opportunities.

Education and training, high value added office services. Some coastal cities and towns are already known as international centres of higher education and training. Others attract 'back office functions' decentralised from large conurbations. The quality of the environment is often the key factor in attracting investment in these economic activities in which a highly skilled workforce is the primary asset. As education and training is a growth industry in its own right, and the flow of office decentralisation from the conurbations is continuing, coastal cities and towns are well placed to secure the investment in these sectors of the economy. Again, attractive coastal urban sites must feature in any 'prospectus' to attract this investment.

Pace of urban change. The third source of economic potential is in the coastal community itself. Healthy cities constantly renew themselves. Prime sites are redeveloped; buildings of historic merit are refurbished. New more valuable uses replace redundant economic activities. Occupiers and investors 'trade up' to more desirable locations and premises. New homes with coastal views entice purchasers from within the town and those seeking a retirement home. Offices in prime locations, and retail centres in attractive settings replace obsolete buildings. By making attractive coastal urban sites available for renewal and redevelopment, planning and development authorities can accelerate the pace of investment.

These three sources of economic growth in coastal are essential to the future prosperity of many communities. Can this potential be tapped?

Releasing the economic potential: promote land use change

In our experience, there is only limited recognition of this economic potential in many development plans for coastal communities. Many local authorities fail to recognise that derelict industrial areas may be wholly inappropriate for modern industrial uses but ideal for residential, tourism and leisure. This problem is not unique to coastal communities as, generally, development plans are slow to reflect long term changes in society's need for fundamental changes in the way land is used.

Similarly, many authorities fail to accept that coastal sites currently zoned for leisure activities, say beachfront promenades and arcades, have slowly deteriorated to an extremely unattractive condition and that urgent action is required.

In other words, a primary constraint on these sources of economic growth in communities is the failure to recognise the need for land use change and the urgent need for investment.

Developed coast NIMBYs are NOMBies

In many communities throughout the UK, the NIMBYs (Not In My Backyard) are an exceptionally powerful force in local politics. (An extreme is Milton Keynes, which was designated to overcome rural NIMBY resistance to providing homes for Londoners. As soon as a family moves to the new city, they become NIMBYs themselves, resenting plans to accommodate other newcomers.)

In our experience, many coastal communities have their own unique brand of NIMBYism. We find that in the political debate on proposed changes of use for urban coastal sites, the NOMBies (Not On My Beach) who oppose any change at all, are a powerful force. We find that NOMBies fall into three categories.

First, there are those who believe that the decay of once grand beachfront promenades, piers, open air unheated swimming pools, amusement parks, arcades and underpromenade arches, and parks is preferred to a sensitive programme or restoration, refurbishment, and redevelopment. Second, some NOMBies believe that any changes to the buildings on the seafront leads to wholesale destruction of the town's architectural heritage and that sensitive redevelopment is, by definition, impossible.

The third type of NOMBie concentrates on opposing the redevelopment of former industrial, warehouse, and other port-related sites for new uses. These NOMBies believe that one day, British industry will again require vast tracts of land adjacent to now redundant dock facilities and that proposals to redevelop sites for housing, tourism, leisure or other uses are undermining the future of the British economy.

NOMBies are extremely effective at opposing ambitious plans to revitalise the developed coast.

Policies to resist pressures or promote opportunities?

As a result of this pressure from the NOMBies most policies for the developed coast in structure and local plans strongly resist change. The draft PPG adopts a similar stance

'... Since the coastal zone is only a small part of the territory of most local authorities, it is reasonable to expect provision of land for housing and employment, for example, to be made elsewhere in the district. Thus, if provision has been made elsewhere, except in exceptional circumstances, policies to resist development on the coastal zone will be supported.'

As far as tourism and leisure is concerned, the draft PPG recognises that these uses 'should normally be guided to existing urban areas, particularly when they can contribute to the regeneration of seaside resorts and waterfront areas.' This stance should be extended to all uses. The key issue, of course, is whether local planning authorities reflect this stance in their policies.

In practice, I believe that if the local planning authority adopts a 'resisting pressure' stance in its policies towards the developed coast, then it will succeed. This is unfortunate as this will hinder urban regeneration and, probably, increase the pressure on the undeveloped coast in many parts of the UK.

The NOMBie philosophy is, of course, appropriate for undeveloped coastal areas but extremely unhelpful for efforts to regenerate coastal sites which are part of existing built-up urban areas. The only way that the regeneration of urban areas, including coastal sites, can be achieved is if a much more positive stance is adopted.

In our experience, there are two main components of this pro-active approach to urban regeneration of the developed coast

(*a*) a visionary development framework
(*b*) public investment in infrastructure, land acquisition and the public realm.

Development frameworks can provide a sound foundation for public and private investment

A local authority or development agency seeking to regenerate key sites on the developed coast can commence the process by preparing a comprehensive development framework. This will include

(*a*) an assessment of the need for the regeneration of the selected developed coast
(*b*) a statement of the vision - in environmental, aesthetic and economic terms - describing the sense of place to be created
(*c*) coastal protection and other environmental requirements
(*d*) main infrastructure requirements for pedestrian and vehicular circulation, car parking, utilities
(*e*) key uses proposed; and zoning of the site for each use so as to identify the main development and refurbishment opportunities
(*f*) public realm landscape strategy
(*g*) urban design guidelines
(*h*) broad phasing strategy and indicative implementation programme including planned disposals
(*i*) capital cost estimates, funding strategy and proposed sources of funds
(*j*) economic and environmental impact assessments.

The development framework describes the vision and how the vision will be achieved. It is not a master plan in that it leaves considerable flexibility for different building design solutions, and often identifies alternative land uses. The framework is a robust skeleton - with the emphasis being on infrastructure, landscape, public realm and implementation - it is not a detailed design.

Preferably, this development framework will be prepared by in-house staff, rather than external consultants. If consultants are used, they should be used to build a joint consultant/in-house team. If possible, the team leader should be a professional who is a permanent staff member with the local authority.

Also, the process of preparing the development framework should actively involve elected members and representatives of the business community, conservation groups, residents associations, and other interests. This should entail more than just consultation - active involvement as well. Project workshops and project steering committees are useful mechanisms. Normally, the process requires between four and six months.

Public investment and development values

The second prime ingredient in a positive approach to the regeneration of developed coast is a stream of public investment in infrastructure, land acquisition, and the public realm. In most UK coastal communities, it will be virtually impossible to promote ambitious urban regeneration schemes without some public sector investment.

The importance of direct access to Treasury capital is being clearly demonstrated by the Urban Development Corporations for the developed coast in Cardiff, Tyne and Wear, and Teesside. Ambitious investments in environmental protection and improvement schemes are being implemented and exceptionally attractive development opportunities are being created for housing, offices, retailing, industry and tourism and leisure.

These developed coast regeneration programmes simply would not have commenced were it not for the availability of Treasury capital. The values of the development sites being created would not be realised rapidly enough to justify private investment in land acquisition, reclamation, infrastructure and environmental protection, and landscape.

In the short run, the costs of preserving, and protecting our environmental heritage will exceed the values created. Yet, in the medium and long term, investments in the regeneration of developed coast are likely to prove to be extremely sound investments. This is demonstrated in the economic appraisals undertaken for most UDC investment programmes.

It is possible for local authorities to take the lead in developed coast regeneration - schemes in Swansea and Hartlepool are excellent examples. Yet efforts to secure the redevelopment of key sites, implement the refurbishment of derelict piers, and make investments in landscaping and infrastructure are severely constrained by restrictions on local authority investment.

The new National Urban Renewal now being proposed by the Government will, hopefully, provide a financial partner for local authorities seeking to promote major developed coast regeneration schemes.

Guidelines for regeneration initiatives
Be ambitious

By world standards, the effort to regenerate Cardiff Bay, with the Barrage at the core of the programme, is an ambitious scheme. The UK Government, via the Cardiff Bay Development Corporation is committed to creating a 'superb maritime environment' which will attract investment from throughout the world.

Similarly, the South Australian State Government and the City of Adelaide are committed to the realisation of plans to build on 2343 ha of low lying, ecologically sensitive, coastal land to attract new investment in the 'core industries' of information technology, telecommunications, environmental management, education and training, health and leisure.

Coastal sites are at the heart of both projects. In a more modest context, equally ambitious opportunities exist in coastal towns and cities throughout the UK. Visionary strategies for the regeneration of Southport or Brighton's seafronts must be seen as vital elements in any attempt to revitalise declining coastal environments.

The key to all of these projects is to define an ambitious vision for the coastal site. This is the starting point for successful coastal regeneration.

Ensure positive local leadership

If at all possible, the local authority must lead the process of building a local consensus around planning policies. Without a positive role being played by the local authority, particularly the elected members, it will be difficult to secure investment from both central Government and private developers. In other words, if there is genuine elected member opposition, or ambivalence, it will be difficult to implement ambitious schemes.

When an Urban Development Corporation has been established, it is still essential to have local authority support for UDC policies. If local authority members oppose the UDC's proposals, considerable delays will result and private investors will be cautious.

Define clear development frameworks in local planning policies

The draft PPG encourages local planning authorities to define the coastal zone for planning purposes and then to define those parts of the zone which should be subject to safeguarding policies, which are suitable for development, and where special conditions on planning consents are required. This is sounds advice.

Yet, the draft PPG is less prescriptive with regard to the developed coast

> '... Plans should seek to improve their attractions as tourist resorts and to regenerate harbour/port areas by designing land as suitable for new functions and activities...'

PRESSURES ON THE ZONE

'Stretches of coast damaged by past industrial, urban development, mining or waste disposal may need proposals for improving the environment. This will be an essential prerequisite for restoring the physical environment and securing economic and physical regeneration. Suitable cases for such treatment will include former industrial sites, port facilities and derelict land. These stretches of coast will often require long term programmes for their regeneration.'

A much more positive approach is desirable; the PPG should strongly encourage the preparation of detailed development frameworks for all key coastal sites in existing urban areas, particularly those which are adjacent to established town centres and/or derelict and underutilised industrial areas. Ideally, the existence of such frameworks should be regarded as a material factor in decisions to refuse applications for development in virgin coastal locations.

This would, of course, require substantial co-operation amongst those authorities with responsibility for a specific coastline. Structure plans could play an important role in specifying the need for such briefs.

A more positive approach by authorities with local plan-making responsibilities to the regeneration of coastal sites in built-up urban areas is essential. This should be used as an integral part of the efforts to resist development on undeveloped coastal sites.

Mobilise pump-priming public investment in environmental protection, infrastructure, landscape, public realm and land acquisition

To secure successful regeneration of the developed coast, the role of the public sector will extend beyond the role of plan-making into implementation. This will entail mobilising pump-priming public investment in key environmental protection works, infrastructure, landscape and the public realm. In the weak property markets that will prevail throughout the 1990s, it is unrealistic to expect private investors to undertake these elements of a regeneration programme.

In addition, in many locations, it will be necessary for public funds to be used to assemble key sites. This will entail the use of compulsory purchase powers, or at least the threat of their use.

In some communities, private developers will purchase the development site at a price which reflects the acquisition, site development and finance costs. In others, some of these costs will have to be written off.

In mobilising this public investment, we will be pursuing medium and long term environmental and economic benefits. It will not be possible, however, to achieve gains by expecting private investors to meet the pump-priming costs.

Be confident that private investment will follow

Most, if not all, ambitious regeneration and redevelopment schemes - both in cities and on undeveloped greenfield sites - entail a leap into the dark. The regeneration of the developed coast is no exception.

There is ample evidence throughout the UK that when very high quality urban waterside environments are created, private developers will seize the opportunity to invest. Sometimes, as in London's Docklands at present, it will take occupiers a little longer to take up the new floorspace or purchase the new units.

Coastal communities still have a special appeal as locations for living, working and recreation. When low and modest risk commercial opportunities to profit from this appeal are offered, residential, commercial and leisure developers will invest. Yet local authorities and, where established, special development agencies, will have to take the lead in creating these opportunities.

Discussion

The importance of comprehensive data sets and on-going monitoring was emphasised.

A general view was expressed that rejuvenation of the developed coast was one method of reducing pressure on the undeveloped coast, although one speaker held that this did not necessarily follow. Public relations were clearly important in the enhancement of developed areas as many NIMBYs (not in my back yard) did not wish to lose the heritage of their recent past. Better schemes were held to have preserved appropriate features and it was suggested that the local plan was the most suitable forum for any battles, for which the PPG should set a positive framework.

Another speaker highlighted the need for agreement by public consultation because confrontation usually implied defeat. Planners and engineers had on occasions in the past blamed each other for mistakes and had to liaise.

The thrust of the paper on planning was to look for a new multidisciplinary mechanism for the coastal area to ensure an overall strategy. All interests had a part to play. The importance of a lead authority was stressed. It was accepted that Coastal Zone Management (CZM) must cross administrative boundaries including low water mark.

The particular problem of protecting some 200 km of railway was mentioned with the need to involve other authorities where the protection had implications beyond the track itself, as at Towyn.

While the need to provide space on the coast to accommodate cyclical change was recognised, most schemes were unfortunately based on existing structures and might even involve a move seaward by the use of beach nourishment.

The examination of 15-20 years' experience in applying CZM in Australia was recommended.

7. Marine nature conservation in the coastal zone

DR SUSAN GUBBAY, Marine Conservation Society

Introduction
The great whales are probably the most powerful symbol of marine nature conservation today. The exploitation of stocks to critical levels, moral indignation over the way they are killed, and the majesty of the creatures have inspired many people to rally to their cause and to marine nature conservation. The hunting of whales still takes place, but public interest in marine nature conservation matters has broadened considerably. Pollution by toxic chemicals, coastal development, nature reserves at sea and many other matters are being given attention alongside interest in safeguarding threatened and vulnerable wildlife. As our knowledge of the sea and our understanding of the impacts which can result from human activity have increased so has the scope of marine nature conservation but it has always lagged behind conservation efforts on land. The contrast is striking. The land area of the UK is around 240,880 sq km with some 242 National Nature Reserves covering 168,107 ha, whereas with around one-third of a million sq km of territorial waters there are only two Marine Nature Reserves which, in total, cover less than 50 sq km. Many reasons are given for this disparity but the three that are most frequently voiced are - a lack of knowledge about the environment and wildlife we are seeking to conserve, less pressure from human activities on the marine environment when compared with land areas, and the difficulties of carrying out the conservation management task at sea. Each of these points is worth exploring to progress marine nature conservation efforts.

Marine nature conservation - the resource
The value of the natural resources of the marine environment are probably impossible to quantify. The tremendous importance of these resources for the well-being of people living in the UK, whether for their financial value or more intangible benefits, is widely recognised. Equally important, however, is the richness and variety of the natural environment of the sea in its own right. As our knowledge of the marine environment has increased so has the appreciation that it is an environment where nature conservation efforts are at least as important and necessary as on land.

Study of the marine environment goes back many years. A 'submarine dive' in the Thames has been reported from as early as 1624 (ref. 1), but the subject developed significantly in the early nineteenth century with the work of natural scientists such as Edward Forbes, Wyville Thomson and William Thompson. Their dredging expeditions to sample seabed flora and fauna laid the basis for the marine biological descriptions of these areas and was complimented by the interest generated in seashore life by the naturalist Philip Gosse. The work of these and others led to early marine biogeographical classifications of the waters around the British Isles. Whilst detailed work still continues on this subject there is general agreement that the British Isles is surrounded by three main marine biogeographical regions. Although predominantly Boreal in nature the position of the British Isles means that Arctic species are found in the northern waters whilst to the south and west Mediterranean/Luistanean species occur quite regularly. The different biogeographic regions reflect an interesting mix of marine species but even more diversity is provided by the varied coastline of the British Isles. This includes the whole spectrum of conditions from extreme exposure to wave action and currents to extreme shelter. When combined with the geology of coastal areas this creates a great variety of marine habitats, with a corresponding diversity of marine communities and species (e.g. ref. 2).

A great deal of work has been, and continues to be, carried out in sampling, identifying, classifying, mapping and trying to learn more about the seas around the British Isles. Oceanographic research in recent years has focussed on the functioning of whole sea environments, geological work carried out by the British Geological Survey has mapped seabed surface sediments around the British Isles and a Marine Nature Conservation Review, started in 1987 by the Nature Conservancy Council (now the Joint Nature Conservation Committee) has been mapping and identifying the marine benthic communities in inshore waters with a view to identifying sites of marine nature conservation value. There is undoubtedly a great deal more to learn about the marine environment but already sufficient knowledge to start identifying sites of marine nature conservation importance. In 1981 the NCC identified seven sites as potential Marine Nature Reserve on the basis of their scientific worth, 77 coastal sites have been identified as qualifying for Special Protection Area status under the 1979 EC Directive for the Conservation of Wild Birds, and in 1986 the Marine Conservation Society published a Coastal Directory which gave information on 264 sites of marine nature conservation interest (ref. 3). In Scotland 29 sites have been identified as Marine Consultation Areas because of their nature conservation importance with a list due out imminently of sites of equivalent value in England and Wales. The research effort needs to continue but the argument that we do not know enough about the natural environment of the sea to identify some important sites and promote their conservation is clearly not valid.

The pressures

Knowledge of the marine environment is undoubtedly the first step for effective nature conservation. This will, for example, help with identifying sites of particular worth for nature conservation. Another important aspect, however, is the threat to the system. The comment that there is a great deal more pressure on resources and space on land than at sea may no longer be true for coastal waters. The expression 'urban sea' (ref. 5), aptly describes some marine areas around Britain yet many people still view the sea as the last great wilderness, relatively unspoilt and largely uninfluenced by man's activity.

The reality of the situation has been brought home, in recent years, with the serious decline in the UK inshore fishing industry, the posting of signs on beaches informing people that it is unadvisable to swim because of a health risk, and parts of the south coast where councils have said there is simply no more room to add moorings for recreational craft. The last example also shows how the situation has got worse over the years. The number of yacht moorings in the Solent has risen from 7,000 in the mid-60s to 30,000 in 1990 and has now reached the stage where Hampshire County Council considers this to be the upper limit which the area can take. This raises the spectre of a sea area being 'full up' for a particular activity - something which would probably not have been considered possible 20 years ago.

The development pressures are also considerable. An RSPB survey of 123 estuaries showed 80 of these at risk and 30 in imminent danger of permanent damage (ref. 6). With around 48% of the British shoreline covered by estuaries this is undoubtedly a significant threat.

There are also growing problems associated with the use of coastal water. More than 300 million gallons of sewage are discharged into coastal waters every day and most of this is untreated. With more than 15,000 km of coastline just 35 sites reach the standards needed to get a European Blue Flag for bathing beaches and there are health risks associated with swimming off traditional beach resorts.

In the North Sea fishing effort is so intense that 40% of the fish biomass is taken out every year; it has been estimated that virtually every part of the seabed of the North Sea is trawled three times a year and the Irish Sea twice a year. This sort of pressure not only affects seabed communities but also the fish stocks themselves. In the North Sea a one-year old haddock has only a 1% chance of reaching maturity at 4 years old. At a time when the world catch of marine fish has been expanding that of the North Sea has stagnated.

The picture presented by these sorts of figures shows that any vision of the seas around Britain as an environment largely uninfluenced by human activity does not match the reality.

COASTAL HERITAGE

Putting nature conservation into practice

The early tradition of nature conservation was to concentrate on individual species with mammals and birds tending to be the focus of attention. This was especially true if their decline could be directly linked to human activity. Conservation campaigns for the blue whale, the oryx and the avocet are some well known examples. The image of a species under threat presented a clear unambiguous message to the public at large - 'your support is needed to help to save this species'. The targeting of individual species for conservation effort still takes place, for example with work on the rhino, the elephant and the tiger, but nowadays it is in a much broader setting. A lot more attention is given to the fact that if individual species are to thrive we must look at the complex systems of which they are a part. Equally a lot more effort is put into explaining these links in conservation campaigns. Nature conservation work has grown from a single species approach to consider habitats, communities and even ecosystems.

In some ways marine nature conservation was ahead of conservation efforts on land in this respect because the single species approach sits uneasily in an environment like the sea. Although attention is given to individual species it soon becomes clear that wider measures are needed. Seal conservation efforts, for example, need to protect the animals, particularly at their breeding and haul out sites, but the health of the seals is also immediately linked to fisheries and water quality. The same can be said of cetaceans, seabirds and fish stocks. The need to take a broader view is even more marked for the vast number and variety of invertebrates in the sea, many of which have some stages of their life cycle in the water column and others on the seabed.

It is soon obvious that just a single species approach to nature conservation cannot work in the marine environment. The targets of our conservation efforts have therefore become the habitats and marine communities - the coral reefs, maerl beds, kelp forests, and horse mussel beds - to name but a few. Even this has its limitations, however, because of a tendency to concentrate on seabed and nearshore communities and largely ignore open water and offshore communities. There is also an assumption that the basic processes which drive the system are continuing to function uninterrupted by human use. This has been taken for granted in the past but the scale of human impact on the marine environment has risen dramatically and may even be leading to major changes to marine ecosystems. The consequences of a greenhouse effect, on a world scale, is a particularly topical example. These large scale changes mean we will need to focus our attention on the healthy functioning of the whole system. This is a most difficult and challenging area of work and one that has hardly been tackled. Marine nature conservation today must nevertheless address the whole range of threats to wildlife and the environment, which means species, habitats, communities and ecosystems and the processes which keep them functioning.

A brief look at the work of conservation organisations shows the tremendous range of measures which are being promoted to achieve nature conservation objectives. Research, education, legislation, and practical management measures - any and all of these can be used to promote specific nature conservation policies, but they must operate against a background of general environmental principles. These must apply to all aspects of human activity regardless of whether it takes place inside or outside a nature reserve, whether it concerns a threatened or common animal or plant, and whether it is a toxic chemical in low or high concentrations. Looking after the wider environment is a key part of succeeding with specific marine conservation initiatives and this is especially true for inshore waters where pressures from human activity are greatest but which also support some of the most productive marine communities. Coastal Zone Management is one way to achieve this because it provides a broad framework in which nature conservation efforts in inshore waters can be successful.

Coastal Zone Management and marine nature conservation

Coastal Zone Management (CZM) is an idea which is increasingly being recognised as a constructive way of tackling problems which result from the pressures and conflicts of use in many coastal areas. A key point is that it is concerned with the whole range of activities taking place in the coastal zone and promotes integrated planning to cope with the management problems which are often found in these areas. Another, equally important, aspect is that CZM is not simply concerned about human use of coastal areas but also the physical and biological environment. Thus CZM does not equate to coastal protection, i.e. the construction of sea walls, groynes etc., but is a much broader idea.

In 1990 the Marine Conservation Society (MCS), in collaboration with the World Wide Fund for Nature, published a report calling for an effective system of CZM for the UK (ref. 7). The report proposed that 'The aim of CZM in the UK should be to ensure the long-term future of the resources of the coastal zone through environmentally sensitive programmes, based on the principle of balanced, sustainable use.' Research carried out for the report showed that there was support for the idea of CZM across a wide range of user groups and that they had a number of common concerns about the current approach to planning and management of activities at the coast. The report recommended eleven areas for action which included the introduction of primary national legislation for CZM, the establishment of a CZM 'unit', the preparation of regional coastal zone plans and the extension of local planning authority jurisdiction beyond low waters.

The benefits of introducing a system of CZM will be widespread and relevant to all coastal users. Its value to nature conservation will also be significant and the following example shows one specific area which will benefit.

Coastal Zone Management and marine protected areas

Many countries have started to safeguard the nature conservation interest of the marine environment by establishing Marine Protected Areas (MPAs). In the UK two sites have been given legal protection: the waters around Lundy Island which lies at the entrance to the Bristol Channel, and the waters around Skomer Island and the Marloes peninsula in Pembrokeshire. The powers to designate these areas were introduced with the 1981 Wildlife & Countryside Act but ten years later only two of the original list of seven proposed sites have been given this status. Unfortunately even in these areas the level of protection is questionable. Part of the problem lies with the legislation itself and part with the procedures for its implementation but it is also true to say that the measures for MPAs in the UK were not put forward as part of a systematic approach to the conservation of sea areas. On land there are a variety of options, for example National Nature Reserves, National Parks, Sites of Special Scientific Interest and Local Reserves. The Marine Conservation Society would like to see a range of options for marine areas as well and is promoting the idea of a two-tier system to benefit as many sea users as possible without compromising the nature conservation importance of a site, for example by actively considering the opportunities to support the interests of fishermen, sailors, swimmers and even underwater archaeologists (ref. 8).

A system of CZM, based on the ideas outlined above will have considerable benefits for MPAs in the UK. Some of these are described below.

(*a*) *Provide a context.* The nature of the marine environment means that MPAs will be influenced and affected to varying extents by activities taking place outside their boundaries. CZM will provide the broader context of management for the sea area outside an MPA and therefore support the conservation efforts within MPAs.

(*b*) *Links with conservation on land.* The landward and marine sides of the coastal zone are interconnected environments yet the boundaries at conservation areas generally perpetuate an artificial divide at 'the coastline'. One of the principles in CZM is that the landward and seaward parts of the coastal zone need to be treated as a unit for planning and management. If introduced this will provide the framework for conservation areas which span the coastal zone and therefore cover more coherent environmental units.

(c) *Providing an overview.* The recommendations of the MCS/WWF report stress the need for a national perspective for CZM. This will give an overview of the status of the coastal environment and the pressures which it faces. The ability to get an overall picture will ensure that the seriousness of issues such as the gradual, accumulative loss of coastal wetlands can be seen more clearly and indeed the broader picture suggests that the integrity of the whole wetland resource of the UK may be under serious threat. The overview clarifies the need and urgency for establishing an effective network of MPAs.

(d) *Public participation.* To be effective CZM must involve the public when both developing and implementing management policies and plans. Public participation is also the key to the success of MPAs yet the current system of consultation for MPAs often encourages confrontation. The Marine Conservation Society has suggested the setting up of a public coastal forum, in each county, for example, as part of the CZM process. This would bring into existence a group who can comment on MPA proposals at all stages and involve concerned individuals much more actively in the process.

(e) *Focus on the coastal zone.* The increased attention to coastal matters, which will come about through the introduction of CZM, will give additional incentive and focus on conservation of this valuable and threatened environment. This will undoubtedly bring benefits in encouraging the establishment of effective MPAs.

Conclusions

Public awareness of the value of the marine environment around the British Isles and the threats to this environment has grown considerably in the last ten years. Unfortunately the statutory options to deliver marine conservation have not kept up with this change although there are a number of initiatives under way. A key idea for progressing marine nature conservation is an effective system of coastal zone management which spans coastal land and the adjacent sea to the limits of UK territorial waters. This will provide the framework in which nature conservation can be successful and will also benefit the interests of other sea users. The need for progress on this issue is pressing in order to move forward with effective nature conservation of the seas around the British Isles - a valuable and threatened environment.

References
1. DEACON M. Scientists and the sea. 1650-1900 a study of marine science. Academic Press, London, 1971.

COASTAL HERITAGE

2. IRISH SEA STUDY GROUP REPORT. The Irish Sea, an environmental review. Part 1. Nature Conservation. Liverpool University Press, 1990.
3. NATURE CONSERVANCY COUNCIL. 13th Annual Report, 1 April - 31st March, 1987. 1987.
4. GUBBAY S. A coastal directory for marine nature conservation. Marine Conservation Society, 1986.
5. SMITH H.D. and LALWANI C.S. The North Sea: sea use management and planning. Cardiff Centre for Marine Law & Policy, UWIST, 1984.
6. ROTHWELL P. Update and threats to UK estuaries. RSPB Conservation Review 3: 28-29, 1989.
7. GUBBAY S. A future for the coast. Proposals for a UK Coastal Zone Management Plan. Marine Conservation Society, 1991.
8. WARREN L. and GUBBAY S. Marine protected areas: a discussion document. Marine Protected Areas Working Group, 1991.

8. The open coastline

C. STEVENS, English Nature

Introduction

Our open coastline, like our estuaries, is one of our great national assets in Britain. On the coasts occur the closest that we have to naturalness and wilderness in lowland Britain, and elsewhere the coasts are a splendidly rich meeting of the uplands and the sea. Their value encompasses the aesthetic, scientific, educational, recreational and economic; they are places where natural forces and systems are dominant, places where man has not subdued nature, and arguably cannot, at least in the longer term. Of course, as an island, coasts are bound to be important but in fact we are gifted with one of the richest coastal heritages in Europe. This paper seeks to summarise this heritage.

Wide, and sometimes confusing, ranges of mechanisms and processes are currently in place as a means of achieving conservation of this area. They are, however, all inescapably locked into the overall management of the coast and can only be successful if our coastal management is itself effective. This paper describes some of the shortcomings of the existing system, and highlights some of the decisions facing us that will have a major impact on nature conservation.

The sedimentary, ecological, hydraulic and climatic systems affecting our coast mean that virtually every part is dependent on, and affected by another. Our coastal heritage is thus far more than the sum of its parts - but to describe it it is helpful to look at the parts one at a time. The major elements are: cliffs, both hard and soft, and the rocky foreshore; shingle and sand ridges, spits, bars, beaches and sand dunes; and saltmarshes and mudflats. These are described in turn below.

Cliffs

Both our hard rocky cliffs and our softer, more erodable cliffs have immense nature conservation value - but there is a major difference in the level of threats, and of man's impact, between these two areas.

Rocky cliffs

Characteristic of the junction between upland landscape and the sea, our rocky cliffs are internationally renowned for their birds, their geology and their scenic value. Prime breeding grounds for seabirds, cliffs from Flamborough north to Dunnet Head, from Cape Wrath to Land's End and the Northern and Western Isles contain the bulk of the European Community's seabirds - and indeed a major proportion of the world population. Over twenty percent of the world's population of razorbill and perhaps thirty percent of the Atlantic form of guillemot nest on our rocky cliffs. Birds can nest safe from predators; the marine larder is at hand - and human interference is less than in any other coastal area.

Many rocky cliffs also have extensive rock foreshore areas. In rock pools and forests of wrack a rich invertebrate fauna finds food and shelter, in turn to sustain birds and larger marine animals.

The very rock of which the cliffs are composed is of great interest too: Britain is often said to have the greatest variation of geological features for its size in the world, and a high proportion of this interest is exposed in rocky cliffs. This is reflected in the number of geological SSSIs, prime teaching localities for undergraduates and for the National Curriculum, designated on our rocky cliffs. Geology is intimately associated with scenic value and it is perhaps in the image of rocky cliffs, dashed by the power of the sea, and topped by untamed uplands that we capture one of the most evocative landscapes in the country.

Erodable cliffs

The contrast between the hard, rocky cliffs and their less stable components in clay, gravel and weak rocks is great. The former have attracted only local development and engineering, while the latter have been a magnet for them. The threat of erosion, the natural location of many soft cliffs in the more densely populated lowland part of Britain where development and pressure on land is greatest and a piecemeal planning system have contributed to huge impacts on this environment - and to building up a legacy of major imbalances in natural erosional and depositional systems (Bournemouth Borough Council, 1989; Herlihy, 1982).

These cliffs are in fact a massive environmental asset and it is essential for our well-being that they are seen as such. Firstly, their erosion is essential to sustain beaches, spits, bars, saltmarshes, mudflats and sand dunes whose value as habitats, amenities underpinning a vast coastal tourism industry, integral parts of our scenery and as elements in our sea defence system are described in the following section (Bird, 1985).

Secondly, the exposures that eroding cliffs provide of our softer rock strata, normally poorly seen inland, is unparalleled in Europe. As a consequence most of the eroding coastline from Flamborough Head south to the North Foreland, and then west to the Exe Estuary, is scheduled as geological SSSI. Such sites include the international reference localities for vast periods of geological time, and have made Britain a major focus, not just for research and education, but also for its close connection with the oil, gas, coal and aggregate industries. Oil industries, for example, use these superb three-dimensional exposures as training grounds for those who must deduce the analagous underground structure of oil fields from the sparce geophysical and borehole data at their disposal.

The third environmental value of cliffs arises from their unique biological habitat. The steady change as they erode landwards allows a unique range of colonising species, notably plants and invertebrates, to thrive. Numerous Red Data Book (the compendium of our most threatened species) species include hoary stock (*Matthiola incaria*) found only on chalk cliffs in southern England (Mitchley and Malloch, 1991) and Cornish heath (*Erica vagaris*) found only on the cliff top heaths of the Lizard.

There are also numerous rare and scarce species, many of them highly specialised. They include the Scottish primrose (*Primula scotica*) on cliff-tops in Orkney and Caithness, and the mason bee (*Osmia xanthomelana*) confined to two sites on the Isle of Wight.

The last, and perhaps most telling, environmental asset represented by eroding cliffs, arises from the fact that they are an extremely powerful, self-regulating, natural system, the disturbance of which (for example, by local prevention of erosion) inevitably concentrates erosive power elsewhere. This concentration occurs not only at the ends of the protection, but also at distances of miles or tens of miles. The fine balance between wave approach angle, sediment transport and coastal orientation, once disturbed, becomes a formidable - and ultimately irresistible - force tending to restore the natural system. The Holderness coast is a prime example. The angle of the coast has been shown to be the product of the eroding regime, and is very precisely in equilibrium. Disturbance by the coast defences at Hornsea has led to greatly accelerated erosion to the south, and hence to a need to extend defences to prevent extremely high rates of loss of agricultural land and, in due course, important industrial installations. This in turn will lead to a need to extend defences southwards again. Here, and elsewhere, a very unpalatable situation is in prospect - a continuous line of defence over vast lengths of coast, maintained at great cost in the face of increasing marine attack, with a background of sea level rise and an inheritance of disrupted beneficial sedimentary processes. The prospect is of a failure of human endeavour in the face of natural forces to rival another crushing loss to the sea, that in the North Atlantic in 1912.

Low lying coasts

As with cliffs there are two contrasting systems that dominate our low lying coasts. The first, generally in the higher wave energy environments, are the shingle and sand ridges, spits, bars, beaches and sand dunes. The

COASTAL HERITAGE

second, generally in the lower energy areas, are the saltmarshes and mudflats.

Shingle and sand ridges, spits, beaches and bars

Shingle and sand ridges and spits act as nature's sea defences on low lying land where wave energy is high. Beaches are the equivalent systems at the foot of eroding cliffs. As with most coastal geomorphological features, they are continuously interacting with the sea, adjusting their profile and position.

Shingle ridges are normally subject to over-topping, often in storms and at spring tides, while spits are known for their periodic breaches. This variability has been traditionally seen by engineers as a weakness. In fact the ability of shingle features to adapt is one of the reasons that they are a major environmental asset. If they are stabilised, or starved of sediment, they lose not only their scenic, wildlife, scientific and amenity value, but here, too, we run the risk of becoming locked into an unwinnable, and costly, battle to hold the line.

Spurn Head is a fine example. The original spit maintained a delicate sediment balance between erosion on its seaward and acretion on its landward shores. A 19th century reclamation scheme in the Humber Estuary severed the sediment supply to the landward shores and the spit practically disappeared within 20 years. Victorian engineering came to the rescue, however, and Spurn was rebuilt using a variety of methods: groynes, rock revetments and sand traps. These have been added to during the 20th century so that Spurn has been held in place by hard engineering for 150 years. During this period the coastal erosion to the north has proceeded, so that the immobile spit has become progressively out of alignment with its mainland 'root'. Maintenance of the groynes and walls has not been kept up, with the result that the spit is now neither maintained by hard defences nor able to maintain itself by natural processes. A catastrophic breach may result, with the loss of one of the most important morphological features on the British coastline. Similar processes are in operation at Orfordness, and at many other spits around the country.

These dynamic coastal sites are also unsatisfactory locations for other coastal developments. Dungeness is an example. It is a shingle spit, and because of its age (it was probably first formed at least six thousand years ago as sea level rose at the end of the Ice Age (Green & McGregor, 1986)) it now has a very rich assemblage of species (Ferry et al., 1990; Philp & McLean, 1985). It is a rich archive for sea level change studies, not only for academic reasons, but also for the very real and pressing task of long-term calibration and validation of the sea level and global warming models, on the basis of which we will presently make far-reaching social and economic decisions. It is also a living laboratory for engineering and scientific attempts to understand the movement of sediments around the coast, an advance that is clearly essential if we are to manage the coast better in the future. Lastly, it is a haven for bird life, with internationally important populations of tern and ringed plover, and it is recognised as a major scenic area.

In this dynamic environment we have built two of the longest term, capital-intensive investments imaginable - Dungeness A and B nuclear power stations. They are built on the tip of the spit, at the area of highest erosion and on the transport route around the end of the spit. We are, as a consequence, now committed to possibly hundreds of years of defending the structure and moving sediment back as it erodes; in the process we have almost completely disrupted the natural systems (Carr, 1988). There may be a range of longer term consequences arising from this, affecting areas beyond Dungeness, but we do not presently know whether this will be the case.

Beaches are the most responsive of all sedimentary systems, often reprofiling themselves in a single tide. Their role in cliff protection is widely accepted, and is exploited using beach recharge schemes. Their amenity value is also well known! The more remote beaches are an important habitat - for example over half of the world's population of sanderling winter on our beaches. They depend on the same invertebrate population that can spoil our picnics, and there is sometimes a conflict of interest when seaweed and other debris in which they live is removed.

As with spits and bars, beaches around Britain have generally deteriorated over the past two decades, and it is widely believed that this is largely a consequence of the inappropriate starvation of sediment arising from coastal protection schemes and harbour developments, as well as other more subtle effects associated with long-term sea level and natural changes in sedimentary systems.

Sand dunes

Sand dunes, like saltmarshes, are storehouses of sediment which also happen to be critically important habitats and to sustain natural coastal systems. Their wildlife importance is massive - for example, dunes contain over ninety percent of our breeding population of the natterjack toad, and five nationally rare or scarce plants and animals that do not occur elsewhere in Britain - like the beautiful dune gentian (*Gentiana ulignosa*) and helliborine (*Epipactis leptochila*). These scarce species are only the tip of a rich and unique ecological system in dunes. Over 120 different characteristic assemblages of vegetation are recognised in dunes (Radley, in prep), making them one of the most varied habitats in the country. Indeed dunes are a major national and international teaching resource for the principles of ecology; the Ynysglas Field Centre, situated against the saltmarshes of the Cardigan coast, is host to many thousands of GCSE, 'A' level and undergraduate students each year - our life scientists of the future.

COASTAL HERITAGE

As storehouses of sediment, sand dunes are cyclic in their erosion and accumulation. It is tempting to react to an eroding dune system by attempting to stabilise it. This is to lose sight of this cyclicity, and of the fact that it is the very instability of dune systems that sustains their wildlife and geomorphological functions. Stabilised dunes lose their value, and it is fruitless to attempt to sustain them to preserve their wildlife and geomorphological characteristics. Dunes do re-accumulate - as the rapid growth of Yarmouth North Denes demonstrates - and sediment lost to erosion immediately sustains beaches, spits and bars elsewhere.

Saltmarshes and mudflats

We can think of saltmarshes and mudflats as being the lower wave energy equivalents of sand dunes and beaches (Hansom, 1988). Saltmarshes have the function of stockpiles of sediment, as do dunes, and mudflats are their source. In engineering terms, beaches and mudflats are wave energy dissipation structures, as are the creeks of saltmarshes (Pethick, in press). The less energetic environment of saltmarshes and mudflats, and the cohesive nature of the material that accumulates there, combine to produce a rich habitat for invertebrates and plants, which in turn provide a food supply and shelter for other animals (Daiber, 1982; Davidson et al., 1991). Saltmarshes and mudflats are in fact the power houses of open coastal and estuarine biological productivity. They also contain rare species - such as the beautiful Essex emerald moth (Waring, 1989).

Saltmarshes and mudflats are also of great geomorphological significance, as natural sea defences capable of absorbing a major proportion of the incident wave energy (Brampton, in press). Indeed, implicit in the design of many of the sea walls built after the 1953 floods, using empirical cross sections, top levels and frontal protection, is the contribution of the then healthy saltmarsh in front of them. As isostatic change in relative land levels in the south-east of the country led to relative sea level rise, marshes have attempted to move inland. This has been prevented in nearly all cases by a sea wall built around the upper saltmarshes when they were reclaimed. Saltmarsh and mudflat profiles have consequently flattened (Pethick, in press) and an alarming number of these (normally accretionary) features are now measurably eroding (Burd, 1992). Their contribution to flood defence structures is being lost, and, perhaps more important, loss of their function as a store of sediment will probably mean permanent loss of sediment from the nearshore zone, making it far more difficult to sustain systems and defences in the future.

This wildlife loss, the threat to the well-being of our natural coast defences and to the economics of manmade sea defences are not the only telling arguments for action to conserve our saltmarshes and mudflats. As reclamation of saltmarsh has taken place piecemeal over a very long period, it is not easy to envisage the extent of loss of saltmarsh. In fact over twenty percent of Essex saltmarsh has been lost in the last fifteen years, and this pattern is thought to be reflected elsewhere, particularly in the south and east of the country (Burd, 1992). This saltmarsh can be looked at in very different ways: as extensive tracts of waste land through a farmer or industrialist's eye, or through an environmental eye as an unspoilt wilderness. Stand looking over the Wash as the sun rises over the North Sea, thousands of acres of saltmarsh and mudflat stretching before you, populated by countless waders and wildfowl - glance a seal's head - and you will get a feel for lowland Britain before man, of the beauty that characterised the saltmarshes before their extensive reclamation. This is one of the few locations where this may be seen in our lowlands, and there is a strong obligation not only to hang on to the remainder of this resource, but also to take whatever opportunities we can to restore as much of it as possible.

Conclusions

We have two coastal heritages. The first is a heritage of powerful natural systems and depleted, but still very important, wildlife. The second is a heritage of piecemeal and often inappropriate coast defence, reclamation and flood protection schemes, many of which will require expensive maintenance in the near future. Sea level rises along the most critical parts of our coasts that have gone on since the end of the Ice Age appear likely to be exacerbated by global warming. This inheritance is going to force us to take an important decision over the next few years and because of the inevitable knock-on consequences of the 'holding the line' approach, this decision amounts to a highly polarised choice between two starkly contrasting alternatives.

The first is to attempt to maintain the present lines of defence. As this paper has attempted to show, this will naturally lead to the prospect of almost continuously engineered structures around all of our erodable and low coasts. With some exceptions this may stretch from Flamborough Head to the Humber, and from there south to the Thames, from the Thames estuary around Kent to Lyme Bay. In addition much of the Cleveland coast and the coast between Sefton and Morecambe is likely to need the same treatment. This would involve virtual complete elimination of the natural features on these coasts, including the loss of their traditional scenery, massive loss of habitat and enormous impact on nationally and internationally important species, together with destruction of large numbers of geological and geomorphological sites. Rising sea levels loom to compound - though not cause - this pressure to engineer. Engineers then will be called upon to stand alone against the forces of nature, without natural systems to assist them, and they are likely to fare as well as their royal predecessor.

The alternative is a two-pronged approach. Firstly, to defend the major features whose social and economic value renders the need for defence beyond reasonable question. Secondly, to accept that natural systems will predominate on the remainder of our coastline. It is a vision of working with - but not giving in to - nature. It shows every sign of making economic sense. It appears more likely to be achievable. Above all it will safeguard one of Britain's leading environmental resources: its natural coastline.

Acknowledgements

The author would like to thank the many biologists, geomorphologists, engineers and planners, both within and outside English Nature, who, through discussion and review, have contributed to the content of this paper.

References

BIRD, E.C.F. 1985. Coastline changes: a global review. Wiley.

BOURNEMOUTH BOROUGH COUNCIL. 1989. Hengistbury Head Management Plan.

BRAMPTON, A. In press. Engineering significance of British saltmarshes. In Allen, J.R.L. and Pye, K. (ed) Saltmarshes, Proceedings of Conference; Reading University, 1991.

BURD, F. 1992. Erosion and vegetation change on the saltmarshes of Essex and North Kent between 1973 and 1988. English Nature, Peterborough.

CARR, A. 1988. Geomorphology and public policy at the coast. In J.M. Hooke (ed) Geomorphology in environmental planning. Wiley, 189-210.

DAIBER, F.C. 1982. Animals of the tidal marsh. Van Nostrand Reinhold Co.

DAVIDSON, N.C. et al. 1991. Nature conservation and estuaries in Great Britain. Nature Conservancy Council, Peterborough.

FERRY, B., LODGE, N. & WATERS, S. 1990. Records and survey in nature conservation, No. 26. Dungeness: A vegetation survey of a shingle beach. Nature Conservancy Council, Peterborough.

GREEN, C.P. & McGREGOR, D.F.M. 1986. Dungeness: A geomorphological assessment. Nature Conservancy Council report 1986.

HANSOM, J.D. 1988. Coasts. Cambridge University Press.

HERLIHY, A.J. 1982. Coast Protection Survey, 19800. Report, 2 vols. Department of the Environment, London.

MITCHLEY, J. & MALLOCH, A.J.C. 1991. Sea cliff management handbook. University of Lancaster, Lancaster.

PETHICK, J. (in press). Saltmarsh geomorphology. In Allen, J.R.L. and Pye, K. (ed) Saltmarshes. Proceedings of Conference; Reading University, 1991.

PHILP, E.G. & McLEAN, I.F.G. 1985. The invertebrate fauna of Dungeness. In Focus on Nature Conservation 12, Dungeness, Ecology and Conservation.
RADLEY, G.P. (in press). The sand dune survey of England. English Nature.
SHIRT, D.B. (ed) 1987. British Red Data Books 2. Insects. Peterborough, Nature Conservancy Council.
WARING, P. 1989. Rescue bid to save the British race of the Essex emeral moth from extinction. Entomologist's Record 101, 231-2.

9. Estuaries

P. I. ROTHWELL, Royal Society for the Protection of Birds

Introduction

Estuaries, the muddy fringe around our shores where rivers and sea meet, are fascinating places with an immense wildlife value. This ribbon of land between the sea and the soil forms part of a biological chain stretching from Canada to Siberia and south to the Cape of Good Hope in South Africa. Millions of birds pass along the ancient migration routes on their twice yearly travels between their winter quarters and their breeding grounds.

Over two million ducks, geese, swans and wading birds depend on the United Kingdom's shores for a winter home and sustenance. Hundreds of thousands more stop over for a few days on their way to and from the breeding grounds. Over 30 per cent of the entire north-west European wintering shorebird population home in on our estuaries like cars to a motorway cafe or aeroplanes to an airport. The estuaries form vast service stations refuelling the birds of many nations. Estuaries are truly international meeting places, supporting wildlife in profusion.

Man's exploitation of his natural world has devastated some of these estuarine stop-overs. Estuarine development and recreational pressure and the possible loss of intertidal land as a result of global sea-level rise give continuing cause for concern. The perils of breeding in the high Arctic and flying many thousands of miles to ensure winter survival set a tough natural test for any bird. Survival can be particularly difficult in adverse weather conditions. The piecemeal development and exploitation of our coast by developers, and the disturbance caused by recreational pursuits, adds measurably to the natural hazards these waterbirds face.

Surveys by the RSPB suggest that, of the 120 or so estuaries surveyed, well over half are threatened by man's activities. More recent work by the late NCC (ref. 1) suggest that this figure is optimistic and that almost all estuaries have been or are being affected by man. Whether or not the effect is a damaging one will vary from estuary to estuary and activity to activity. Some are in imminent danger of large scale permanent damage; the threats are many and varied.

The sand and silt flats of our estuaries may seem vast, but their capacity to absorb further damaging development is limited. The loss of seemingly small areas can threaten the integrity of the entire system. Even modest sized developments can have a disproportionately large impact on birds. The

RSPB believes that urgent Government action is required to safeguard what remains of this irreplaceable habitat.

Biological value of estuaries

Estuaries are among the most biologically productive systems in the world. The intertidal muds and silts support many invertebrate species, often in large numbers. Wading birds in particular have evolved to exploit this rich resource. We are fortunate in the United Kingdom, owing to our long coastline intersected by rivers great and small, in having a wealth of estuarine habitats. These sheltered shores, bathed by a mix of nutrient-rich waters from river catchments and saline water in shallow seas, have a range of habitat types from submerged mudflats, intertidal mud and sand through to saltmarsh. They support an abundant array of wildlife. In addition to migrant and wintering waterbirds, estuaries provide spawning and nursery grounds for fish, including many of considerable commercial value.

Estuaries extend around much of the United Kingdom and southern North Sea coasts. Large tidal ranges, warm seas, mild winters and a heavily indented shore combine to provide conditions highly suitable for the formation of estuaries. Despite the UK's relatively northern latitude, our coastline generally remains ice-free in most winters, allowing birds access to the intertidal feeding grounds throughout the winter. Our coasts thus provide the closest wintering grounds for migrants from high Arctic breeding areas.

The UK holds very important areas of saltmarsh, the higher zones of which can be rich in plant and invertebrate species. England's east coast saltmarshes form 26 per cent of the total area of saltmarsh around the North Sea (ref. 2).

For breeding birds, saltmarshes are most significant as a habitat for redshank. Out of a total UK breeding population estimated at 35,000 pairs, about 20,000 breed on coastal marsh (ref. 3). Other species, such as gulls, terns, lapwing, shelduck and many others, also utilise saltmarshes.

The main ornithological interest of the UK's estuaries is their vast population of wintering and migrant birds. Birds come from breeding grounds in temperate, Arctic and sub-Arctic areas from northern Canada in the west to central Siberia in the east, across Greenland, Iceland, northern Scandinavia and Russia. The UK is at the centre of this migration route. For migratory wildfowl and wading birds, the western European estuaries form vital links in a chain of sites used for moulting and for replenishing fat and muscle lost during long migration flights. Many species stop briefly in the UK and rest here en route to wintering grounds farther south. Local migration movements do occur, particularly by wildfowl in hard weather, but many waders are remarkably site faithful and return to the same places year after year (ref. 4).

Wading birds and wildfowl have received a considerable amount of attention both here and abroad. Co-ordinated counts (refs 5 and 10) show that

(a) in January 1990 a peak total of 1.7 million wading birds was recorded by the Birds of Estuaries Enquiry; this accounts for 42 per cent of the total population in the whole of north-west Europe in winter
(b) 41 UK estuaries each hold over 10,000 wading birds in winter
(c) at least 36 UK estuaries hold regularly over 20,000 waterfowl (wading birds and wildfowl) and thus qualify for recognition under the Ramsar Convention for Protection of Wetlands of International Importance
(d) wetlands in the UK are considered to be of 'national importance' if they hold over one per cent of the population of one species or subspecies of waterfowl; over 60 estuaries in the UK meet this criterion and thus are recognised as nationally important.

For individual species, the UK is particularly notable for

(a) 87% of north-west Europe's knot spend the winter on UK estuaries
(b) the Wash alone supports one quarter of the population of knot which breed in Greenland and Canada
(c) other species of note include

oystercatcher	over 33% of the north-west Europe population
grey plover	over 61% of the north-west Europe population
dunlin	over 41% of the north-west Europe population
bar-tailed godwit	over 35% of the north-west Europe population
redshank	over 73% of the north-west Europe population

These figures do not include the large transient populations of birds on passage that migrate onward to other countries after visiting the UK.

Turning to wildfowl, the UK species of note include

shelduck	over 31% of the north-west Europe population
wigeon	over 32% of the north-west Europe population
teal	over 22% of the north-west Europe population
pintail	over 33% of the north-west Europe population
light-bellied brent geese	circa 75% of the north-east Canada/north Greenland population
dark-bellied brent geese	circa 50% of the world population
barnacle geese	circa 34% of the world population of which 13% (the entire Svalbard/Spitzbergen population) are present on the Solway Estuary
pink-footed geese	over 30% of the world population can be found on the Ribble estuary in some winters.

These vast and important numbers of birds are by no means evenly spread around our coasts. Individual sites can achieve extremely high values for certain populations and species.

A joint RSPB/NCC exercise has identified, using carefully selected criteria, species for which Britain is either 'internationally important' or which are 'vulnerable' through being rare or localised, or which have recently suffered considerable declines in range or numbers. These birds are listed on the UK red data list (ref. 6). UK estuaries support 20 red data bird species for some part of their life cycle.

The International Council for Bird Preservation (ICBP) and the International Waterfowl and Wetlands Research Bureau (IWRB) have produced an inventory of the most Important Bird Areas in Europe (ref. 7). Data assembled by the NCC, RSPB, BTO, WWT and others led to the listing of 261 IBA sites in the UK, of which 58 are estuaries.

We can thus conclude that estuaries provide an internationally important UK wildlife resource.

Estuaries also play a fundamental role in the economic system of coastal states. Seven out of every 10 people in the world live within 80 kilometres of the coast, and almost half the world's cities with populations of over one million are near estuaries.

Estuaries are the most productive part of the marine ecosystem. Coastal and inshore waters are a vital source of food, yielding all but 10 per cent of the world's fishing catch. All plaice and sole caught in the North Sea grow up in estuaries. This nursery function gives estuaries great economic importance. Annually, the North Sea yields 2.6 million tonnes of fish. The North Sea is the most profitable sea in the world. Much of this productivity depends on a healthy environment. This immense and profitable biological productivity is, however, being put at increasing risk by other human activities such as pollution, over fishing and land claim.

The threats

Estuaries have long been subject to interference from man. The Romans were active in claiming land for agriculture in the Thames Estuary. As our technology developed, so the ability of man to alter the natural state of affairs in estuaries increased. The number of threats, their widespread nature and the difficulty in predicting the effects of piecemeal development on birdlife constitutes a complex problem.

The results of a recent survey of threats carried out by the RSPB in the UK gives a guide to the areas of greatest concern (ref. 5).

Of the 123 estuaries surveyed (80 per cent of the UK total), 80 were considered to be under some degree of threat, with 30 estuaries in imminent danger of permanent damage.

The Nature Conservancy Council has gathered data on the extent of knowledge of the estuarine habitats of England, Scotland and Wales and the activities affecting them (ref. 1). The initial results of NCC's survey indicate that the RSPB has under-estimated the level of threat to estuaries.

Some developments, notably marinas, barrages, port expansion, land claim and industrial uses, are irreversible and reduce the extent of intertidal habitats available to birds. Others, such as bait-digging or cockle fishing, have more temporary impacts. If properly controlled, the damage they cause can be minimised, and is mostly also relatively short term and reversible. Collectively, however, they can deny large areas of suitable habitat to birds.

Habitat loss has been strongly implicated in the steep decline of at least one wader species, the dunlin (ref. 8). In this instance the loss has been attributed to the spread of *Spartina*, an invasive saltmarsh grass, as well as development of the upper shore, and consequent loss of mudflat. Similar losses are occurring in the south-east of Britain, where natural subsidence of the land, and the possibility of rising sea-levels, will also reduce the extent of intertidal land.

Land claim is widespread, it has affected at least 85% of British estuaries and has removed over 25% of intertidal land from many estuaries (ref. 1).

For some sites, the battles have been largely lost. The intertidal area of the Tees Estuary has been reduced by around 90 per cent in the last 100 years, having been lost to land claim for port and industry related developments.

In another north-eastern England site - the Tyne - all the intertidal flats have been lost to land claim. In Northern Ireland, 85 per cent of the intertidal area in Belfast Lough has been the subject of industrial development and landfill.

'On just 18 estuaries at least 89,000 ha have been claimed - 37% of their former area and overall an almost 25% loss from the British resource. In the last 200 years estuarine land has been claimed at 0.2-0.7% per year.' NCC 1991 (ref. 1).

Widespread land claim is continuing. The NCC work confirms that 123 land claims in progress in 1989 affected 45 estuaries. Land claim was for rubbish and spoil disposal, transport schemes, housing and car parks, and marinas.

Much historical land claim created coastal grazing marshes of substantial conservation value. This grazing marsh has often now been secondarily claimed for intensive agriculture and for urban and industrial developments. Between 30% and 70% of the marshes that remained in the 1930s in different parts of south-east England have now been claimed (ref. 1).

International wildlife value does not reduce land-claim pressure. 26 estuaries in the Ramsar/SPA network have current habitat losses, 21 of these losing intertidal/subtidal aress (ref. 1).

Land-claim proposals also affect 36 Ramsar/SPA estuaries. Over half of our internationally important estuaries face land-claim proposals both piecemeal and major.

Protection
Sites of Special Scientific Interest (SSSI)
(Areas (ASSI) in N. Ireland)

The main conservation legislation procedure in Great Britain is the Wildlife and Countryside Acts 1981 and 1985. Under their aegis, the successor bodies to the late Nature Conservancy Council have a duty to identify and designate SSSIs. The importance of this designation is recognised in Government circulars which indicate a presumption against damaging development, but no binding constraint. The SSSI procedures require landowners and occupiers to consult with the NCC when activities that might damage the site are being considered.

In Northern Ireland, the DoE (NI) administers the equivalent designation - Areas of Special Scientific Interest (ASSI).

Problems with the existing SSSI/ASSI system limit its effectiveness in a number of ways, especially the following.

(a) A number of internationally important sites have not yet been notified as SSSI/ASSIs and may therefore be damaged by commercial or recreational activities.

(b) Although most activities which occur on estuaries notified as SSSI/ASSIs are controlled by the Wildlife and Countryside Act, some potentially damaging operations are not. In particular, those governed by common rights, planning controls and private Acts of Parliament specifically override the protection afforded by the SSSI notification.

(c) The geographical coverage of estuaries by SSSIs in England and Wales extends only to median low water, thus excluding and failing to give any protection to those areas falling between this point and low water spring tide. This area is often considerable in size and can be important for birds. In Scotland, however, SSSIs do extend to the extreme low spring tide mark. In Northern Ireland, no decision has yet been made as to the lower tidal limit of ASSIs.

Problems also occur where the high tide refuges for birds are located on habitats which are not of any other scientific value (e.g. arable land adjacent to the estuary). Damage to or disturbance of these important areas can affect bird populations and should be covered by the designation and management process. The SSSI designation does not in itself prevent damage and loss to wildlife interest. In 1988-89 the NCC reported 241 cases of damage to SSSIs including 22 damaged by activities that had received planning consents.

Internationally Important Estuaries: Ramsar Sites and Special Protection Areas (SPAs)

Ramsar Sites. The Ramsar Convention on the Conservation of Wetlands of International Importance was ratified by the UK Government in 1976. In doing so, the UK accepted responsibility to promote the conservation of wetlands of international significance within its territory. In respect of birds, this is defined as a site holding over 20,000 waterfowl, or regularly supporting 1 per cent of the individuals of a flyway population of one species or sub-species of waterfowl.

It is explicitly stated in the Convention that there is an obligation for the contracting parties to include wetland conservation considerations within their national land-use planning. They are required to formulate and implement this planning so as to promote the wise use of wetlands in their territory. Significantly, the contracting parties have interpreted this wise use requirement to mean the maintenance of the ecological character of wetlands. At least 36 estuaries in the UK meet the criteria for Ramsar designation.

Special Protection Areas. Under the European Community Directive on the Conservation of Wild Birds, adopted in April 1979 (79/409 EEC), the UK Government is required to take special measures to conserve the habitats of two categories of bird. These are the rare or vulnerable species listed in Annex 1 of the Directive, and all regularly occurring migratory species.

Member states are required to designate suitable areas as Special Protection Areas (SPAs) and to protect these from damaging developments. Attention is particularly drawn to wetlands, especially those of international importance.

In Britain, the NCC is responsible for identifying sites suitable for designation as Ramsar sites and as Special Protection Areas. The UK Government requires that before any area in Great Britain or Northern Ireland can be considered for designation the area should be an SSSI (or an ASSI). The NCC has identified 150 candidate Ramsar sites and 216 candidate Special Protection Areas throughout Britain. Of these, 55 contain estuarine habitats. Sixteen of these have been wholly or partly designated as Ramsar sites, and 14 are also designated SPAs. In Northern Ireland, the Government's responsibilities rest with DoE (NI), but no SPAs and no coastal Ramsar sites have been designated in the province as yet.

Progress towards listing has been painfully slow, and has, if anything, slowed down in recent years. The UK now lags behind other EC member states in the area of land protected as SPAs.

Designation as a Special Protection Area confers an additional level of presumed safeguard on the site. The UK Government has advised local planning authorities that planning permission for a development should, as a rule, not be granted within an SPA or a candidate SPA. Development should be allowed only where the disturbance to birds or damage to habitats will not be significant in terms of the survival and reproduction of the species.

Not only has progress in making designations to safeguard our wildlife heritage been lamentably slow, but the designations themselves once applied have also proven to be no guarantee of protection. The provisions of private Acts of Parliament (through the Private Bill system) and planning consents still lead to damaging impacts. The many candidate sites awaiting designation under the Ramsar Convention/EC Directive are particularly vulnerable. Thus the UK Government is manifestly failing to protect sites of international value. This must be addressed if the UK is not to lose credibility in the field of international wildlife conservation.

It is insufficient merely to label internationally important sites as such, without then ensuring that they are protected by the application of good management practice. The designation should be seen as the starting point for wise management and sustainable use. The concept must be encouraged by suitable resource provision to management bodies (such as NCC) to implement the estuary action plan. Local authorities must be made more aware of the importance of the Ramsar/SPA designation. It is an outstanding accolade conferred on a UK nature conservation site and should be seen and appreciated as such.

Estuary conservation

The RSPB believes that a concerted programme of action must be undertaken to safeguard the UK's estuaries and the birds and other wildlife that depend upon them (ref. 11). New measures are required to protect our coastal environment, coupled with an upgrading of the value of estuaries in the eyes of decision makers. Above all, a new attitude to the needs of the internationally important bird populations found on our soft coasts must be encouraged.

Strong site safeguard measures are the key to protecting intertidal habitats. Action must therefore be taken by national and local government authorities to address the needs of estuaries and the international heritage of the UK's waterbirds.

Currently, over 33 Government departments, statutory bodies and agencies have responsibilities in coastal areas. Add to this local authorities, land owners, managers and users, and the scope for confusion and duplication of roles becomes clear.

To effect the strong site safeguard measures and to guide local planning authorities Government must bring order to this confusion of bodies with powers to affect the coastal environment.

Coastal zone planning

Many of the development pressures and threats to estuaries stem from inadequate protection through our own political processes. Recognition of the ornithological value of estuarine sites through statutory designation alone is no guarantee of protection. On estuaries, the Dee estuary still suffers from pollution, disturbance and tipping, despite its recognised national and international importance. The Orwell, Medway, and Severn estuaries, among others, are similarly being damaged by activities with planning or parliamentary approval.

If the objective of protection of all nationally significant sites is to be met, Government must plan for their wise use within a national strategic planning framework.

At present, the combination of piecemeal planning and lack of an overall national plan for the coastal zone, exacerbated by a multiplicity of different bodies with jurisdiction in coastal areas, gives rise to confusion and poor application of policy.

The Department of the Environment should play a co-ordinating role in the management of our coastline and inshore waters. There is a need to review strategic guidance in the maintenance and management of estuarine and coastal habitats. There is a clear need for a Government department with major UK-wide remit for controlling the management of our coastal zone. A standing committee of statutory interest chaired by the DoE is needed as a first step to improve interdepartmental liaison. A body charged with the co-ordination of coastal management should be identified or created.

The lack of attention to wise management of estuaries, following recognition and designation as a Ramsar site or SPA, betrays the attitude of Government to such international responsibilities. The designations are often seen as an end in themselves, rather than as the start of a 'wise management' process. Often they are regarded as preventing development (including benign development).

The broader view, however, would emphasise that the nationally and internationally important sites are part of a continuity of interest around our coasts. The safeguard of this system overall can be achieved only by comprehensive strategic control of damaging developments and other activities, and the implementation of adequate management schemes for our most important sites.

The management of the biological value of estuaries should be seen as a logical part of a national planning process implemented at local level.

Estuarine systems have traditionally catered for a wide variety of uses. This has led to the development of the multi-layered system of jurisdiction that is so inappropriate to effective and co-ordinated coastal zone management.

Conclusion

The international nature of the wildlife interest of the UK's estuaries has long been undervalued by both the general public and government. The immense habitat losses and the apparent unwillingness of government to address its international responsibilities are testimony to a lack of appreciation. Domestic wildlife legislation is demonstrably inadequate in the protection of one of our most important wildlife resources. International convention and law similarly would appear to be limited in its application in the UK. A prerequisite to the development of management strategies for estuaries is the acknowledgement of the international value to the resource.

Man has long been meddling in the estuary ecosystem and in terms of impacts we can identify two main types

>   (a) firstly the impact of land take - dock and harbour building, extension and management, agricultural, waste disposal and industrial development - this clearly involves permanent habitat loss which has an obvious impact on birds
>   (b) secondly, man has traditionally used estuaries for many practices that might be termed temporary and reversible in their effects: fisheries activities, bait digging, wildfowling and recreation all fall into this category - if stopped or managed effectively their impact is minimised and their continuance can be assured in a manner commensurate with sustaining ecological character.

This scenario logically leads towards two main conclusions. Given the undoubted international conservation value of the estuaries of the UK there should be

>   (a) no further estuarial land claim
>   (b) adequate management to ensure the sustainability of estuarine wildlife value.

References
1. DAVIDSON et al. Nature conservation and estuaries in Great Britain. NCC, Peterborough, 1991.
2. BURD F. The Saltmarsh Survey of Great Britain 'An inventory of British saltmarshes', Research and Survey in nature conservation. NCC, Peterborough, 1989.
3. ALLPORT G. et al. Survey of Redshank and other breeding birds on saltmarshes in Britain, 1985. Unpubl. RSPB report to NCC (includes references to other surveys of saltmarsh breeding birds). 1986.
4. PRATER A.J. Estuary birds of Britain and Ireland. Poyner, Calton, 1981.

5. SALMON D.G. et al. Wildfowl and wader counts 1989-1990. Wildfowl and Wetlands Trust, Slimbridge, 1990.
6. BIBBY C. et al. Towards a Bird Conservation Strategy. Conservation Review, No. 3, RSPB, Sandy, 1989.
7. GRIMMETT R.F.A. and JONES T.A. Important bird areas in Europe. ICBP/IWWR3, Cambridge, 1989.
8. GOSS-CUSTARD J.D. and MOSER M.E. Rates of change in numbers of dunlin, *Calidris alpina*, wintering in British estuaries in relation to the spread of *Spartina anglica*. J. Appl. Ecol. 25, 95-109, 1988.
9. WELLS J. Waste disposal and conservation in Northern Ireland. RSPB internal report. RSPB, Belfast, 1988.
10. SMIT C.J. and PIERSMA T. Numbers, midwinter distribution, and migration of wader populations using the East Atlantic Flyway. pp 24-36 in H.Boyd and J.Y.Pirot (eds). Flyways and reserve networks for water birds. IWRB special publication 9. 1989.
11. ROTHWELL P.R. and HOUSDEN S.D. Turning the Tide - a future for estuaries. RSPB, Sandy. 1990.

Discussion

The practical outcome of the NRA, Anglian Region, strategic approach to North Norfolk was eagerly awaited. The short-term approach must not be allowed to prejudice long-term policy.

An example was given of a Norfolk conservation partnership. There was a need for conservation bodies to speak with one voice. Co-operation on conservation was taking place between MAFF, the NRA and English Nature.

One speaker suggested that there was a need to recreate mud-flats, although some doubt was cast on the likely outcome. The Australian experience at Botany Bay was mentioned as an example of birds developing a new habitat after being displaced.

While it was accepted that the defences at Hornsea inevitably interfered with the coastal processes along the Holderness coast, the practicality of abandoning the town in the long term was questioned. The possibility of continuous defences from Flamborough Head to the Humber was not considered to be a reality.

A policy of retreat was not necessarily a panacea as it was a move from a certain to an uncertain scenario. A requirement for a roll-back researcher was mentioned.

10. Development pressures in an area of rising human resources

M. N. T. COTTELL, FEng, FICE, FIHT, MBIM, Consultant, Travers Morgan

Oh, I do like to be beside the seaside
Oh, I do like to be beside the sea.

Introduction
For decades, if not centuries, men and women have yearned to be near the sea. Whether for work, home, leisure or industry, as a nation, we have always loved the sea. In a country with 15,000 km of coastline and where no one is more than 135 km from the sea, the attraction of the sea, the need to use the sea and the influence of our coastline is possibly more significant than for any other nation on earth. The trade of past generations depended upon our ability to import and export across the oceans of the world. That is changing - the sea link to mainland Europe will not be the only form of crossing from mid 1993, when the Channel Tunnel opens. Nevertheless, as an island, we will continue to depend to a large extent upon the sea for our livelihood.

Some businesses gain profit from the sea itself - fishing, oil and aggregates, and many endeavours of work or leisure depend upon the sea for transportation or for sporting activity. With modern travel, few have not experienced the joy or discomfort of being on or near the sea. Indeed, many of the resorts established some 100 to 200 years ago would not have developed but for the lure of the sea.

History of development
Development has taken place at the boundary between sea and land, the coastal strip, for a whole series of reasons, but the two major pressures over the last two centuries have been the need to use the sea for our livelihood and the pleasure - almost all enjoy being "beside the seaside".

In the past, a further development pressure arose from the need to defend our nation against our enemies. While construction for defence may now be unnecessary, development of our coastline by the army, navy and airforce have changed the coastal zone we now need to manage.

DEVELOPMENT PRESSURES

Because of its proximity to continental Europe, the South East of England has seen many developments on its coastline for defence purposes. The martello towers of the Napoleonic wars are significant, but the major castles often have their foundations in earlier history; some date to the days of Norman conquest, while Dover Castle ramparts date to the prehistoric Iron Age. However, the use of our coasts for defensive purposes became acutely necessary in the Second World War, when many defensive fortresses played a major role in "our darkest hour". During this period, the traditional defences were added to with pillbox fortifications and the use of cliff top for radar and airfields.

As a maritime nation, our ports and resorts constantly evolve. There are many places of major development which once enjoyed the tranquility of a fishing village. Brighton's Lanes - much frequented by tourists, were originally fishing village streets. Also development comes and goes, like an ebbtide, over many decades. Our coastal areas have a number of towns and villages which were thriving ports at one time, but due to siltation, changes in cargo handling, the tonnage of ships and the available landward transportation, have declined or changed their role, often becoming tourist attractions.

All these activities have taken place over many decades; indeed, the development of the coastal zone for fishing and defence has evolved over centuries.

While housing development has followed the defence and fishing industries, it was not until the impact of modern transportation systems, initially railways and, at a later date, roads, that people were able to live by the sea and work elsewhere. Indeed, an early advertising campaign for BR - Southern Railways - stressed the joy of living on the Sussex Coast and travelling to work in London by train. Some coastal towns such as Herne Bay were created by the advent of the train and the ability of people to work in London and live near the sea comparatively cheaply.

However, the development of the seaside resort in Victorian times led to further development pressures, and while this took place in most parts of the United Kingdom, the South Coast towns today create an almost continuous development from Ipswich to Margate and onward to Bournemouth. A combination of easy access to London and holidays by the sea created this pressure for development.

The transformation in later years of these traditional resort towns along our coasts has created further development pressures as the attitudes to holidays by the sea have changed dramatically since the Second World War.

The resort towns have had to diversify into conference and service industry activity to provide jobs for the resident population; indeed, some also encourage manufacturing. The growth of Europe as a community and the importance of London as a financial centre have continued to create development pressures in the South East and, in particular, along its coastlines.

Moreover, the decline of the defence establishments has presented new opportunities for the reuse of land for other developments. Some of the South Coast airports of today would not have received planning permission had the area not been used as a wartime airfield. Many Second World War military developments have been used to establish industrial and tourist industry locations, many South Coast caravan parks being located where the army or air force provided roads and hardstandings. An excellent example of effective reuse is Chatham Naval Dockyard, closed in 1984. This area of 210 hectares has been dramatically changed, with part becoming an historic trust, part used as a RoRo container port, and a major area being redeveloped for housing and industry.

The growth of trade with Europe, and the need for new port facilities to handle the modern ferries and their passengers and freight, have created further pressures on the South East of England. Not only is Dover the largest ferry port in the world, but also there are some 33 ports from Ipswich to Bournemouth. While some are specialist ports dealing with one or two commodities, all these ports have development for freight storage and handling, and the change of transport mode at a port usually encourages other industrial activity nearby.

The Institution of Civil Engineers' publication "Infrastructure - the challenge of 1992" draws attention to the problems of ports. It comments that South Eastern ports are growing fast and are well integrated into the pattern of container movements, but the road and rail facilities serving them are not integrated into the EC system. Access to ports is a fundamental part of the economic health of the country and Government can ensure appropriate investment.

A further development that requires water is power generation. Over the last 50 years, most power stations have been located adjacent to major water sources and with the power needs of the South East, many stations, of necessity, have been located in coastal areas. While new ideas of electrical generation may change the concepts of "big is beautiful", the transport of cheap coal from overseas is likely to sustain many of these power stations for some years to come. The Kent coalfield has also created its own coastal development and pressure for reuse of industrial land now the mines have closed.

A further development between the two world wars was created by the "leisure plots" concept. At Whitstable, the Isle of Sheppey and Peacehaven, these developments have had a particular effect upon our coastline - not only adding to the development sprawl, but often created on coastal cliffs or alluvial areas where the problems of erosion and inundation are high risks for the property owner.

DEVELOPMENT PRESSURES

In recent years, development pressures on coastal areas continue. The increase in population is one factor, as is the drift towards London and the South East. The constraint policies on development in London and the green belt force many to commute into the city each day. People who work in London and live beyond the green belt often choose a coastal resort as the place to buy their home.

Fortunately, since the Second World War, effective planning control has limited development on the few remaining areas of natural coastline. In recent years the need to conserve these areas of countryside and unspoilt coastline has been intensified, and a number of agencies have varying responsibilities for the conservation of our coastal areas. They include English Nature, RSPB, National Trust and English Heritage, to name but a few. The creation of Heritage Coast, SSSI or ANOB status helps planners and developers alike to understand the importance of these few remaining areas of natural habitat and unspoilt coast and countryside.

Development pressures are not just new buildings; the increased leisure time and spending money available to almost all has created further problems on these coastal areas by walkers, holiday makers and weekend trippers, often creating an unacceptable burden on the coastal area.

Recent development examples

The modern physical developments and the conflict with preserving countryside and coastal zone in South East England are characterised by the Channel Tunnel terminal; an area of land between the M20 and the North Downs escarpment has been used for this purpose; also the tunnel spoil has been deposited in manmade lagoons at Shakespeare Cliff - major changes to part of Kent's landscape.

Developments at Sheerness are also significant, with major port expansion and land reclamation on marshland coveted by bird lovers. A new container terminal on the Isle of Grain is located on land which formed part of the now disused BP oil refinery site, adding extra pressure to the Medway Estuary and the road and rail communications on the Isle of Grain.

Ramsgate was developed as a RoRo and passenger port linked primarily to Dunkirk. While certain undertakings were given about the scale of this development until a new road access was achieved, these became difficult to enforce, when the proposed roadway was refused planning permission following a public inquiry. Initially strongly supported by the district council, because of the need to create jobs, the County Council negotiated a development formula with the present owners until the new road could be built. After investigating eleven options and public consultation, a new route along the shore from Pegwell Bay to the Harbour was put forward. In practical terms, this route was the only cost-effective solution which removed dock traffic from Ramsgate's historic streets. The scheme was also to incorporate a new coast protection wall and promenade. An inquiry in 1987 heard evidence from many objectors and the decision to refuse planning permission was based on three main points

(a) the presence of rare algae in caves at the base of the cliff
(b) uncertainty about the port's future after the Channel Tunnel opens
(c) predictions of traffic growth put forward by the port operators.

This example demonstrates the frailty of the planning system. The perverse nature of this decision can be highlighted by "people versus algae", as the cliff erosion will threaten development in the years to come and some 900 heavy lorries continue to pass through the historic town of Ramsgate to this day. Indeed, the questioned growth in traffic through Ramsgate put forward by the port in evidence at the public inquiry, has been more than borne out in practice. It seems likely that the port will survive the competition of the Channel Tunnel with the ever-increasing movement of people and goods to continental Europe.

While evidence at a public inquiry can always be questioned at the time, and afterwards, a fault of current legislation is the difficult task anyone, be they individual or local authority, has in questioning a decision of the Secretary of State following an inquiry. As a nation, we have clearly failed to appreciate the needs of development and the impact on our coasts and the traffic on road and rail. We continue to under-invest in the basic infrastructure needed to support our economic wellbeing as a partner in Europe.

The Ramsgate Harbour access road and coast protection wall may be raised again before the prospect of local government reorganisation stops sensible investment for a 5 to 10 year period, but to reverse a previous decision by the Secretary of State will require a major effort, new evidence and goodwill from Government. Faced with some opposition to such a scheme, it seems likely that Ramsgate will have to tolerate heavy lorries for a long time and the erosion of the cliffs will continue.

The planning regime

After years of experience nationwide, the current systems of controlling excessive development while allowing development in some areas are well rehearsed. Moreover, planning authorities are asked in guidance notes "to bear in mind particularly that land protected from tidal flooding by artificial defences could be extremely vunerable in the event of breaches". While the National Rivers Authority are always consulted on plans within such areas, the planning authority does not have to follow any advice. Indeed, there are reports of planning authorities allowing development without taking account of the coastal problems that could occur.

DEVELOPMENT PRESSURES

A planning policy guidance note specially on the coastal area is being contemplated by the government. The county structure plans will almost certainly contain reference to coastal area problems, although they may not be specific statements, being an integral part of either policies on settlements or countryside.

Kent's structure plan clearly aims for the expansion of economic activity and employment through the growth of existing industry and commerce and by attracting new firms, but within a series of policies related to conserving the countryside and coast. As the Channel Tunnel commences operations, a number of jobs will be lost in the ferry industry as they streamline their operations to compete. Current planning policies allow for development in the coastal towns to offset this loss of jobs. The structure plan also supports port development at all existing ports. However, new development for tourists, such as holiday chalets and caravan parks, on undeveloped coast will normally be refused. The structure plan is in favour, however, of tourist attractions and facilities.

Structure plans change with time, and it's likely that more attention will be paid to recognising the conflict between nature conservation and tourism, and to promote positive management of the coastal zone to try to ensure a balance is achieved. Kent is also likely to define its Heritage coast and a management action plan. But this plan will be about conservation and planning policies. It will not contain all the essential features of a coastal zone management plan. It is likely that the next structure plan, in consultation with the NRA, will deter housing in areas at risk from tidal flooding.

Local plans, prepared by district councils, give more specific definition. Some such plans contain policies on new building in coastal areas, generally prohibiting development in coast protection zones, where cliffs and coastal slopes are subject to erosion. For areas likely to be subject to tidal flooding, some plans emphasise making adequate means of escape in the interests of the safety of the house occupants. This seems a curious policy in an area where adequate development land exists elsewhere, in view of distress caused to householders when their property is ruined by tidal inundation.

Some port and marina developments lie beyond the scope of local planning control and may well require an Act of Parliament before they are passed or refused. However, other developments in ports can proceed without reference to local authority or parliament because they are development of operational land. These developments can be significant in terms of coastal management as they may include areas of sea and beach nearby.

Coastal management

In planning any major development, the recognised conservation agencies (NT, EH etc.) are consulted, and their views are usually an important consideration in any subsequent planning decision. The question that does not seem to be adequately addressed, relates directly to the theme of this conference "Coastal Management". Before a development is approved in a coastal area, who determines the risks to such development from tidal flooding? Who controls or determines the need for coast protection, and its scale? In other areas, who advises on the stability of cliffs or shore, where are the experts on littoral drift, harbour siltation, the need for and effect of dredging etc? In the creation of new ports or marinas or their extensions, who defines the acceptability of the increased shipping and boating that may follow?

The problem is not that all these experts do not exist - they do - indeed the skills of the British planner, engineer, hydrologist, oceanographer etc. are as high as found anywhere in the world. The problem is that often the developer and the local authority, while consulting widely, do not always understand the forces of the sea and the potential problems that can result from a lack of a full appraisal of the coastal regime.

Hampshire County Council have endeavoured to address this problem by the production of "A Strategy for Hampshire's Coast" (ref. 3). In this document, Hampshire presses for a change in the legislative and management of the coastal zone seeking

(a) jurisidiction of planning control to extend to 3 miles from the shore

(b) changes in legislation relating to coast protection and sea defence; administrative responsibilities to secure a more integrated approach to coastal planning and management

(c) discussions with other agencies on how the co-ordination or strategic management can be improved. While this is a praiseworthy initiative, it also highlights the need for new legislation and management arrangements which a county council cannot enforce.

The ICE report (ref. 1) reminds its readers that all sea defence and coast protection works have to be justified on technical, economic, social and environmental grounds and that the value of land or built environment which may be lost to the sea is an important criterion in the decision process. It promotes the formation of "coastal cell groups" for the entire coast in order to achieve comprehensive coastal zone management plans. While some areas of the coast have such groupings many areas do not. And while the ICE has established the Coastal Engineering Research Advisory Committee (CERAC) and the Coastal Engineering Advisory Panel (CEAP) and Government has formed the Co-ordinating Committee on Marine Science and Technology (CCMST), there is still a need to manage the coastline more comprehensively.

There are suggestions that the European Community are to address this administrative problem by way of a directive. While this is clearly a move in

DEVELOPMENT PRESSURES

the right direction, and one cannot prejudge its thoroughness in tackling the problem, it nevertheless requires local understanding of cause and effect before real progress will be made. There is a need for coastal management engineers to ensure they fulfil a wider role and perhaps encourage a better understanding by discourse and publication.

The House of Commons Environment Committee has been studying the issues of coastal zone protection and planning, and the Institution has established a UK Coastal Engineering Data Base and produced a report on Future Sea Levels. The Institution's statistical survey of storm damage to coastal defences in the winter of 89/90 shows that over 110 km were damaged, 18 km of seawall, 12 km of revetment, 9 km of sea banks, 22 km of dunes and 16 km of cliffs, not to mention the damage caused as a consequence to roads, land and property. In Towyn alone, the breach of the sea wall flooded 10 sq km of low lying land, damaging 2800 homes and 6000 caravans. The local authorities and British Rail may face prosecution. If a case is brought it is likely to highlight the unco-ordinated way the coastline is managed.

The lack of proper understanding of the roles of Local Authorities, County Councils, Government, NRA and other bodies was addressed in a Green Paper published in 1985. Unfortunately, the discussion on this document was delayed due to privatisation and restructuring of the Water Authorities.

The option for the UK must now be to develop a new legal framework of responsibilities, incorporating a new and positive approach to managing the coastal zones. As a nation, we must plan, engineer and execute professional coastal management and the responsibilites for such work need to be clearly articulated in new legislation. With the close relationship to the physical planning of development now operated by county and district councils, we need to consider the best approach to more positive action plans. These interactions of coastal management and planning development are particularly acute in areas of development growth such as the South East of England.

However, the Government has established a commission (ref. 4) to look at the future of local government. Most commentators seem to believe this will result in single all-purpose local authorities similar to the solution adopted in 1986 in the Metropolitan areas, when County Councils were scrapped. However, the shire counties and districts present different problems. Most services are executed by the County Councils, but districts, being the local unit of government, often have a better working relationship with Parliament through their MPs. The 1986 reorganisation established joint boards for some services in the metropolitan areas, and these have been questioned since, due to their lack of accountability to the electorate.

It is not the purpose of this paper to debate the future of local government, but if unitary authorities are the favoured solution, it seems likely that they will have to be larger than most shire districts if the problems of joint boards are to be avoided.

The future restructuring of local government could, however, be a further reason for delay to the establishment of proper coastal zone management.

What are the essential features of coastal zone management? Clearly, the coastal zone needs definition. It must include foreshore and near foreshore, any areas where construction may affect the regime of the shoreline, the need to protect coastline or where cliff stability is in question. The zone should also include areas affected by coast erosion or possible threat of inundation.

Coastal management needs two complementary organisations: a national authority which could be a government department, or an authority such as the NRA, to develop a national strategy, provide guidance, scientific support and national information and maintain national data bases. Such an authority will, almost certainly, have some responsibility for securing at least part of the funds required, and would allocate funds to the local management unit.

Some will argue that the local function of developing detailed local plans and operation, inspection and construction should be executed on the basis of the regional coastal groups already formed in some areas. However, this approach creates large organisations which are "Quangos", where local knowledge of conditions may not be available due to their scale of activity. Such a solution also creates difficulties for the expenditure which should be locally funded and denies true accountability to the electorate.

If the results of the local government commission are reasonably populous unitary authorities, there is a strong argument for these authorities assuming the positive coastal zone operational management, because they will have considerable local knowledge, provide the local finance, prepare both structure and local plans and are accountable to local people. These authorities also have responsibilities for people when emergencies occur, will prepare emergency plans and co-ordinate action in the event of a disaster caused by flooding or erosion.

Using the new local authorities under the umbrella of a Department of State, where all the responsibilities are brought together, would clarify the responsibilities of management, ensure a proper basis for funding and that the structure and development plans of an authority are drawn up in close liaison with the coastal zone management authorities who would be accountable to the same electorate. It would also guarantee that one authority is responsible in any emergency.

Conclusion

This paper endeavours to explain the historical and current development pressures in the coastal zone, to identify the current methods of planning control and the inadequate arrangements for appraising development applications in coastal areas. Past experience may have often been right, but there is a sufficient number of examples to make professionals of all disciplines concerned about the failures that have occurred and the need to put matters on a firmer management basis. With this in mind, the establishment of coastal zone authorities based on the future local government structure is proposed, so that true accountability for action and expenditure to the local people is maintained.

References
1. Infrastructure - the challenge of '92. Institution of Civil Engineers, 1992.
2. The planning strategy for Kent. Kent County Council Planning Department, May 1990.
3. A strategy for Hampshire's coast. Hampshire County Council Planning Department, June 1991.
4. Local Government Committee for England. Draft Guidance, paras. 38 and 39.

11. Development pressures in an area of declining human resources

M. J. HORSLEY, Assistant Director of Planning, Cleveland County Council

Background

The River Tees and its estuary serve what was once recognised as one of the nation's important growth points in North East England. One of the country's most important centres of the growing iron, steel and engineering industries in the nineteenth century, it added shipbuilding, chemicals and petro-chemicals to its heavy capital industrial mix in the twentieth century. Hence population, a mere 8,000 in 1801 (Middlesbrough only 25) had already increased to about 40,000 by 1851, 361,000 by 1921 and 570,000 (Cleveland County Council) by 1981.

By the early 1960s, the Teesside and Hartlepool area was a designated growth zone in the Government's Hailsham Report and the subsequent White Paper (ref. 1) for the North East where it was stated

> "70. But the Committee recommended that any very substantial new development should be on the Tees, and that priority should be given to the necessary work there.
> "71. In the Government's view this recommendation fits in well with the conception of Teesside as one of the main centres of expansion for the region as a whole... the Minister of Transport has invited the Chairman of the new National Ports Council to arrange that when the question of the future port development on the Tees is considered, account will be taken not only of the stimulus that improved port facilities can give to Teesside's growth but also of the prospects of increased economic activity in Teesside itself arising from the Government's policies..."

Indeed, this concept of Teesside as a growth area was built into the Northern Economic Planning Council's report (ref. 2) published by the Department of Economic Affairs in 1966 where it stated

> "10(a) The Growing National Population - We believe that the region has the capacity to absorb part of the growing national population. We ask the Government: (i) to consider Teesside as a national and not merely a regional growth area..."

DEVELOPMENT PRESSURES

"107 Teesside - including the area from Hartlepool to Darlington - is the area with probably the greater long term development potential and the survey now being carried out there will help its rational development. It is one of the most economically favoured parts of the region, with ample level land, not subject to subsidence, capable of development, capacity for considerable port development, relative freedom from dereliction ... a commercially strategic position reinforced by good lines of communication to the south by air, rail and road."

The reference to a survey, commonly known as Teesplan (ref. 3), set the scene for the joint structure plan exercise between the newly formed Teesside County Borough, Durham County and North Riding County Councils in the early 1970s. The emerging structure plans had to rationalise the enormous growth forecasts, originally set out in Teesplan, from an overall population of Cleveland (excluding the Hartlepool area), of about 700,000 in 1991 to just under 600,000. At the Examination in Public this was further reduced with a forecast of 526,330 being accepted as the most realistic estimate for 1991. These changes came about because of the dramatic decrease in jobs in the area and increased the net outward migration already being experienced in the mid 1970s (ref. 4).

Since 1971 Cleveland's population has declined and the current estimate stands at 550,000 for 1991. Since this estimate includes 90,000 in Hartlepool, one can deduce that even with those adjustments, population is some 60-70,000 short of the forecasts made seventeen years ago. One might conclude therefore that the area is not the growth centre which was being forecast in the 1960s and that it is an area of declining human resources. However, as you will see later in this paper, in land use terms the area remains dynamic and is still a very important developing area in regional and national terms. The fact that it has had to face up to all kinds of new development issues has maintained the focus of attention from a very wide range of organisations. Changes in its administration have resulted in experimentation and change in all kinds of ways. While the demand for growth is still there and pressures for development appear to continue, I believe there is a genuine desire to co-operate and ensure that it is not at the cost of the environment. Even so, there remain areas of continuing conflict and difficulty in realising an acceptable development pattern which meets all interests. It is the local planning authorities' task to see that a proper balance is achieved.

In the preparation of this paper I checked back on another interesting source of information from the past and that was a report of The Coasts of North East England (ref. 5). This report followed a coastal regional conference - one of a series arranged by the National Parks Commission in 1966 following the Ministry of Housing and Local Government's publication of

Circular 7/66 on The Coast. As ever, the Tees Estuary lying along the boundary of the two regions, seemed to be relatively ignored considering its importance. There seemed to be a failure to identify the interelationship between rivers (especially significant estuaries) and coasts and more attention was paid to recreation and amenity than to nature conservation and pollution which are key issues in the 1990s in terms of sustainable development and acceptable development. Furthermore it is interesting to note that the then County Planning Officer of Durham is quoted as saying, with regard to the coast just to the north of Cleveland where, as he said, his own authority had for years been worried about the level of employment and industrial activity in the County and where one or two industrial sites had been allocated on the coast for 10 to 15 years, "If someone came along and said he would dearly like to develop industry at Blackhall where unemployment sometimes reaches 8% - I would have to be very bold to reply, 'Let unemployment stay at 8%: we must keep this bit of coastline as it is'." That was the 'practical reality' which faced planning officers every week and much as he "cared about the coast of Durham and the north east, I would not pretend to give them priority", he said.

It is interesting to note that because of the decline of population and jobs the greater has been the temptation to accept development in areas of high unemployment (and Cleveland's unemployment in November 1991 was 14% and had reached 22% in November 1985). However, I do believe we have turned a corner in this respect. Policies to protect the environment, both for its own sake and in the recognition that investment and jobs are more likely to come to a physically attractive area, figure strongly in today's plans for the Tees Estuary and Cleveland Coast.

Figures 1 and 2 show the River Tees and its estuary as it was 150 years ago with the position in the 1980s. A vast reclamation programme has taken place, narrowing and straightening the course of the river and replacing the mud flats with flat land, with the result that nearly 6,000 acres has been reduced to about 400 acres of valuable feeding areas for migratory and wintering waders and wildfowl. Clearly this has, to a significant degree, affected the sand and dune coasts to the north and south of the estuary, the quality of both river and coastal waters and the general amenity and attractiveness of the area. This paper does not describe in any great detail the coastal policies and management arrangements but it is worth noting that to the south-east of Saltburn, some eleven miles from Teesmouth, the coast changes dramatically where the magnesium limestone cliffs rise sharply to a height of about 600 ft (at Boulby). This is the commencement of the 33 mile long Cleveland and North Yorkshire Heritage Coast which runs down to Scarborough, broken only where the Skinningrove steel works is situated and around the developed areas of Whitby. The Heritage Coast forms the North Sea boundary of the North York Moors National Park, whose northern

DEVELOPMENT PRESSURES

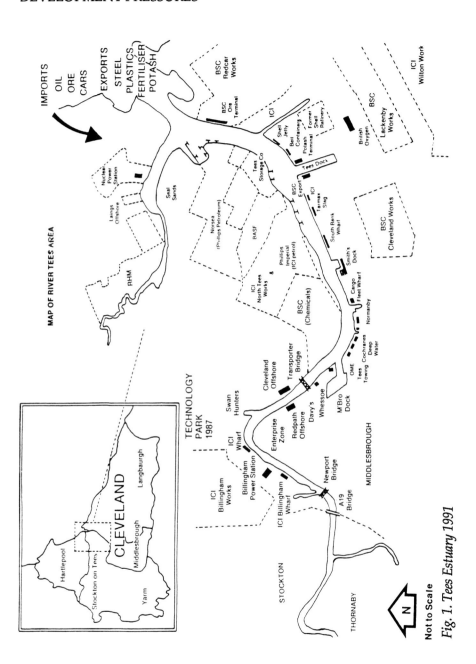

Fig. 1. Tees Estuary 1991

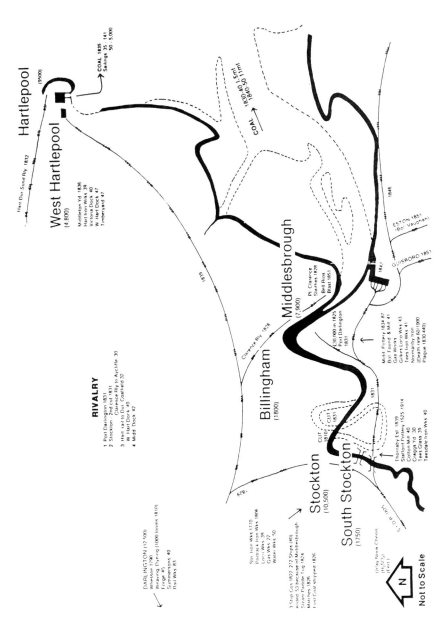

Fig. 2. Tees Estuary, early 19th century: 1810-50 enterprise

boundary offers dramatic views north over the flat lands of the Cleveland conurbation and Tees Estuary. So outstandingly beautiful areas are visible and readily accessible from Teesside.

Development pressures

Development pressures on Teesside will, I imagine, be very similar to those of other major estuaries. They may be divided into the following four categories

(a) physical and structural alterations which can affect the shape and geography of the river and estuary

(b) changing patterns of communication and transport and related infrastructure

(c) related patterns of land use development and water area activities

(d) issues related to environment and amenity

(e) developing organisations and their inter-play and influence.

Physical and structural changes

Figures 1 and 2 show the degree to which the River Tees has been reshaped. Between Stockton and Middlesbrough two major cuts in the river shortened its navigational length considerably in the early part of the 19th Century (Fig. 2) and left large areas of flat, poorly drained land. This area remained open until recently with the exception of one or two developments close to Thornaby and Middlesbrough for housing and industry, the extensive railway marshalling yards between the A66 and River Tees a golf course and race course. Old tracks have been replaced by modern dual carriageway trunk roads and the crossing point of the A66 and realigned A19 meant that such a central open area was always under pressure for development.

The most dramatic changes have taken place, downstream of Middlesbrough around the estuary of Teesmouth. Railway lines built along the south bank of the river involved substantial reclamation between the lines and the river channel. The need to safeguard the area from flooding resulted in a series of Tees Conservancy Acts (ref. 6) which gave the Tees and Hartlepool Port Authority and its predecessors enormous powers of reclamation for port related development, much of which did not go through the rigorous planning procedures required since the 1970s. It was under these powers that the major reshaping of Teesmouth took place with the reclamation of Seal Sands. This, and the extraction of brine, and the development of ICI Billingham as a major chemical plant, helped lay the foundation for a changed Teesside, widening its economic base from the 1920s onwards and stimulating the post-war petro-chemical industry developments. While this is all history, these development pressures set up a momentum and pattern of development which, once established, still have enormous implications for the area.

During the 1975 Examination in Public of the area's first structure plans the issue of the reclamation of the final 420 acres of Seal Sands for port and port related development was a major topic for discussion. The Secretary of State for the Environment's conclusion in 1977 went contrary to the Panel's advice (ref. 4). He stated that, if justified in the national and regional interests, plans should make provision for the potential need for deep water facilities by making this land available for port and port related industry. He then, in recognition of its ornithological interest, added all kinds of qualifications to maintain the area undisturbed as long as possible. He instructed the County Council to investigate the creation of an alternative facility to fulfil at least some of the educational and scientific interest of the area, recognising that the reclamation of Seal Sands would result in a major loss of such interests (ref. 7). Consequently various studies were carried out in the late 1970s and early 1980s but solutions were found to be technically poor substitutes and practically impossible because of the lack of co-operation of land owners (ref. 8).

It is worth noting that there are current proposals to radically alter the nature of the River Tees upstream of Middlesbrough. The Teesside Development Corporation obtained a private Act of Parliament last year which allows the construction of a River Tees Barrage (ref. 9). This will have the effect of reducing the tidal length of the river by approximately 10 miles, maintaining the level of water upstream at a constant high level. This is intended to make the river cleaner upstream of the barrage, more attractive for new development along its banks (notably a major urban area renewal scheme on former industrial land at Teesdale including canals, boat moorings and a university). Clearly there are questions about its long term hydrological effects on the more polluted stretches of the river downstream to Teesmouth and on drainage and bridge structures upstream, but ultimately the scheme is intended to be of major benefit to the river and its appearance and is in accordance with the River Tees Plan (ref. 10) and Cleveland Structure Plan (ref. 11).

As development has moved downstream and industrial plant and port activities concentrated towards Teesmouth, the need to provide better highways and service roads along both banks, but especially to the south of the river and the need for new crossing points of the river, has become more urgent. At the same time development on reclaimed derelict land upstream, between Stockton and Middlesbrough is more intensive than was originally planned for, especially since the Teesside Development Corporation's rapid progress on their flagship schemes, Teesdale and Teesside Park.

In the late 1970s and early 1980s two major bridges were constructed over the River Tees: a high level three lane dual-carriageway structure for the

DEVELOPMENT PRESSURES

north-south A19 trunk road when it was diverted from its old route through Yarm/Eaglescliffe/Stockton and the new A66 trunk road at Thornaby. Currently three more road bridges are planned upstream of the barrage: a north/south bridge to serve the new Thornaby Bypass, and two linking Stockton to the new TDC Teesdale development, one of which crosses the barrage.

The relevance of this to the Tees Estuary and Coast is that, as with all coastal areas, there tends to be a major break in coastal communications because of the complexity of crossing such wide estuaries. That is particularly so in Cleveland where Hartlepool and Redcar appear to be relatively close, but in fact the lowest crossing points of the River Tees are at Middlesbrough, the last two being somewhat antique and unique structures - the Newport Bridge and the only remaining operational Transporter Bridge.

The Teesside Structure Plan, 1977 originally proposed a tunnel linking South Bank, Eston to Seal Sands. However, its only justification was one of more closely integrating the new Cleveland County and its traffic contribution was always questionable. Consequently, with the decline in population and development it was deleted as Structure Plan policy by an alteration in 1983, there being only one objection received.

Apart from new bridges and roads the whole Tees Corridor has been the subject of a massive grant aid programme related to the reclamation of derelict land and replacement of the full infrastructure, including new sewerage schemes, drainage and replacement of basic gas, electricity and water services (ref. 12). The Tees Corridor programme expenditure profile between 1984 and 1988 was around £36 million, with about £18.5 of that coming in the form of EEC grants. It was initiatives such as this which attracted the Government to the idea of setting up a Teesside Development Corporation. Derelict land reclamation schemes between 1974 and 1989 involved a further £3.75 million using Derelict Land Grant aid and the County Council's River Tees Environmental Improvement small schemes involved a further £913,000 together with an annual expenditure of £30,000 on the Teesdale Way and Tees Valley Warden Service. This demonstrates the great efforts that are being made to improve the river. While much of this work is upstream from the estuary it does also include many local improvements around Teesmouth to improve public access and vantage points, especially at the North and South Gares.

Patterns of land use and development around Teesmouth

From what has already been said it will be understood that there has been a movement gradually downstream to the north of the River Tees. Since the war, development around Teesmouth has been significant. It has included the Hartlepool Gas Cooled Nuclear Power Station (constructed 1970-84) and the Philips Petroleum Refinery and Oil Storage complex which received planning permission as the land upon which it was sited was being reclaimed with a mixture of slag and dredged landfill. The ICI Philips Imperial refinery, BASF and other 'notifiable' chemical plants on the north bank opposite Tees Dock were also constructed in the 1960s (Middlesbrough Dock was closed in 1983).

One of the most controversial planning decisions in the early 1970s was the approval in 1974 of the British Steel plant at Warrenby on the coast between the South Gare and Redcar. Originally this was seen as a means of not only retaining steel-making on Teesside, but also was intended to bring an extra 10,000 jobs into the industry at a time when steel plants were closing down all over the place. The original proposals involved 4 phases. Phase 3 would have involved building the plant substantially out to sea below the current low water mark. However, phases 3 and 4 were never approved. It may well be that the decision did help to retain the steel industry in Cleveland because as this paper is being completed the closure of Ravenscraig in Scotland has just been announced with possible implications for Cleveland. The proposal involved one of the country's first significant Environmental Assessments and much soul-searching because of its effect on the coast and nearby sites of scientific interest (ref. 13). This is a good example of the sort of pressures local planning authorities have to meet and face up to in an area of industrial decline.

The other good example has been the degree to which Seal Sands should be reclaimed. I have mentioned earlier the Tees and Hartlepool bid to have the whole area reclaimed for industry and port development in the early 1970s and the Secretary of State's decision in 1977. This followed a substantial exercise involving a wide range of bodies first considering the implications of handling the Philips Refinery application, then studying the implications of North Sea oil and rig construction (there is rig construction yard at Graythorp next to the Hartlepool Power Station). From this an informal North Tees Master Plan was produced which aimed to strike a balance between industry and the environment. However, because of the Structure Plan controversy and decision this was never finally approved. The pressures of the 1970s relaxed somewhat during the 1980s and the impact of North Sea oil was never as great as had been anticipated; consequently we were all given some breathing space and the effect of this was already reflected in the decisions taken to resist more port development by the Secretary of State when he approved the Hartlepool Structure Plan in 1980.

Most recently development pressure has come again from the North Sea. This has resulted in approvals for the landing of pipelines and treatment plants for receiving and processing natural gas from the Central Area Transmission System (CATS) involving new gas of pipelines between Redcar, Wilton, Seal Sands and Billingham. Apart from serving the new Teesside Power Station (ENRON/ICI) at Wilton, a 1875 MW gas turbine generating

DEVELOPMENT PRESSURES

station, speculation that other power stations might seek to develop valuable industrial land is a major concern. Associated with such developments is short-term disturbance of some SSSIs and the long-term implications of a whole new network of 400 kV overhead electricity lines which do not improve the environmental image of the downstream area. Again the attraction of the scheme is cheap and clean electric power generation and steam for ICI Wilton and the possible saving of jobs there as a consequence.

Finally improvements to the River and the reduction of industry in some of the old port areas of Middlesbrough, Stockton and Hartlepool have created a desire to introduce more attractive leisure, recreation and housing schemes in such areas. The Hartlepool Marina development is one of the TDC's major flagship schemes and this is matched in scale by the Teesdale development at Stockton. The National Rivers Authority has recently embarked on a joint study of the recreational uses of the River. Similarly the coastal communities of Seaton and Redcar are seeking to upgrade their image as resorts by various improvements, some involving City Challenge bids.

Environment and amenity

Perhaps the most dramatic bit of news to focus on this paper is the agreement in 1989 between the Port Authority the Nature Conservancy Council (now English Nature) and the Crown Commissioners to protect the remaining part of Seal Sands (ref. 14) (see Fig. 3). Essentially this granted to the NCC a fifty year lease for the management and enhancement of the Seal Sands and Seaton Dunes and Common SSSIs in return for allowing industrial development of two small portions of relatively unimportant conservation value. The final decision depended on the Secretary of State agreeing the proposals as part of his decision on the Cleveland Structure Plan in 1990, which he did while at the same time commending the agreement (ref. 15).

The Seal Sands area is part of a complex of SSSIs at Teesmouth and the Tees estuary is one of at least 29 estuaries along the North Sea Coast which are internationally important for wildfowl and waders (ref. 16). While not as important perhaps as the Humber and Wash it does support the passage of a substantial population of wader species and meets the criteria for designation under the Ramsar Convention (Wetlands of International Importance). One of the results of reduced employment and industrial development has been the removal of capital-intensive industrial land allocations between Billingham and Seal Sands. The Structure Plan acknowledges this as an area for development of an International Nature Reserve which is one of the 'flagship' proposals of the Teesside UDC.

The other important questions of environment relate to the nearby concentrations of industry and pollution levels along both the River Tees and the coast of Cleveland. Clearly there is the impact of EEC Bathing Beach

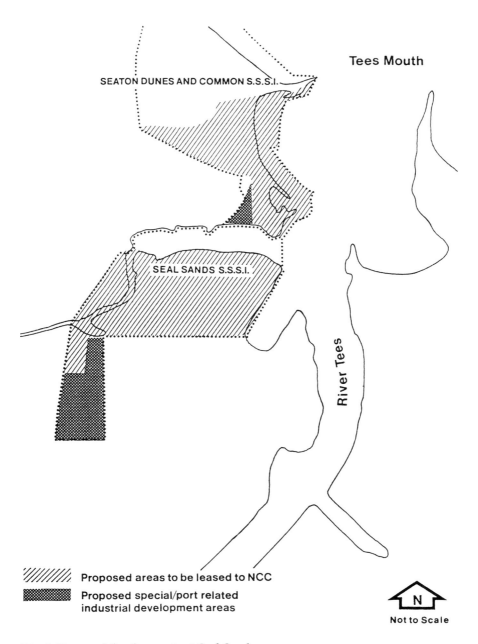

Fig. 3. Proposed developments at Seal Sands

DEVELOPMENT PRESSURES

Standards and programmes are underway for full treatment of sewage as well as the construction of a long sea outfall at Seaton. To the south of Teesmouth a long sea outfall and partial treatment scheme were completed Redcar/Marske in the 1980s to link to inland sewerage systems. The Portrack Sewage Works, again constructed in the 1980s, depends presently on the dumping of sewage sludge at sea. Since this is likely to be ended in the 1990s questions arise as to how to deal with this problem. Considerable arisings of industrial toxic wastes occur in Teesside and there have been two applications in 1989 refused for Rotary Kiln Industrial Waste Incinerators which have been considered at a six month linked public inquiry, the results of which are awaited.

As regards the quality of the river water the former Northumbrian Water Authority introduced a phased programme of improvements which the National Rivers Authority is responsible for achieving. The construction of the Barrage, together with more complete sewage treatment should make a major improvement upstream but there are still questions about the enforcement of consents and the impact of the capital intensive industry concentrations around Teesmouth.

Finally it should be noted that many of today's proposals require Environmental Assessments involving considerably more time and expertise and discussion with consultants.

Organisational structures

Teesside has been subject to a large number of local government structural changes - Hartlepool 1967, the Teesside Borough Council 1968, Cleveland County Council and four new Borough Councils 1974 and now we are once again in the throes of local government reorganisation discussions.

The area has been subject to considerable and extensive forms of experimentation and change, with some schemes relating to the river and coastal areas. I refer to

(*a*) Enterprize Zones - Hartlepool and Middlesbrough
(*b*) Inner Urban area programmes
(*c*) the largest Urban Development Corporation in the country with huge investment programmes
(*d*) changes in the running and organisation of the port authority culminating in its recent privatisation proposals
(*e*) the privatisation of British Steel
(*f*) the privatisation of electric power generation and distribution
(*g*) changes in the organisation of nature conservation bodies, in particular the setting up of INCA (Industry and Conservation Association)
(*h*) powerful lobbying interests beyond the more traditional industry and union lobbies.

I could go on, and each one is a story in itself. However, the overall situation is one of constant change and in my experience it is amazing how one or two key personalities can stand out in this process and influence changes.

Surprisingly out of all these changes, some of the guidance local planning authorities have been seeking has not been there from the governments of the day. For example, where was a national ports strategy when we wished to identify the issues of port expansion or contraction? Where was the national guidance on the protection of estuaries and coast? Very often estuaries and coast issues have to be tackled in a vacuum.

Conclusions

Although Teesside/Cleveland is an 'area of declining human resources' there is a great deal of development pressure and change taking place in the area around the Estuary and coast.

My experience tells me not to accept all proposals on face value. They often involve considerable in-depth examination of all environment, legal and political issues.

Infrastructure development, such as the Tees Barrage or the Marina at Hartlepool, takes a long time to emerge as practical schemes, but I have seen sufficient of these on Teesside not to give up hope - what is needed is plenty of patience and cash.

Patient one may be, and without much cash, but local authorities have to sell their case for improvements all the time and to do this they must be sensitive to and build on local community concerns. This process involves constant and close contact and networking with the lobbying bodies.

Finally I have ducked the issue of sustainable development. However, I believe most of what I have said relates to the issue of a constant drive to get the right balance between economic needs and growth and protecting and enhancing the environment. That is what planning is all about. I believe that success in that process is essential on Teesside to attract development and investment of the right type, otherwise the area could lose the very resource on which it is based. The River Tees should not be regarded as just a working river but as a major attraction of the area.

References

1. The North East - A programme for regional development and growth. November 1963, HMSO Cmd 2206, Page 24, para.70 & 71.
2. Challenge of the Changing North, A Preliminary Study. Department of Economic Affairs, HMSO, 1966, Page 4 (para. 10(a)) and 71 (para 107).

DEVELOPMENT PRESSURES

3. WILSON H. et al. Teesside Survey and Plan. Final Report to the Steering Committee, Vol.1, Policies and Proposals, HMSO, London, 1969; Vo1.2, Analysis, HMSO, London, 1971.
4. REPORT OF PANEL. Teesside, East and West Cleveland Structure Plans, Examination in Public, June/July 1975, Pages 5-20 and Pages 29-30.
5. NATIONAL PARKS COMMISSION. Coastal Preservation and Development. The coasts of north-east England. HMSO. London, 1968.
6. Tees and Conservancy Acts of 1852 to 1919 and 1920. Power to Execute Works, Sections 4 & 5.
7. SECRETARY OF STATE FOR THE ENVIRONMENT. Notice of Approval Town and Country Planning Act 1971 (as amended) by TSCP Act (1972) Teesside Structure Plan, Paras 8.5 and 8.6, 31 October 1977.
8. DAVIDSON N.C. Seal Sands Feasibility Study. A report to Cleveland County Council and the Nature Conservancy Council. May 1980.
9. TEESSIDE DEVELOPMENT CORPORATION. River Tees Barrage and Crossings Act 1990.
10. CLEVELAND COUNTY COUNCIL. River Tees Plan for Recreation and Amenity (adopted 1981).
11. CLEVELAND COUNTY COUNCIL. Cleveland Structure Plan 1990, Policy Inf. 3, Page 60.
12. THE TEES CORRIDOR - County of Cleveland, European Development Fund. A Programme for Regeneration 1985-87.
13. British Steel Corporation. Proposed complex - Redcar phases II and III - Report of the Teesside CBC Chief Officers, 1972.
14. HEADS OF AGREEMENT. Nature Conservancy Council, Crown Estates Commissioners and the Tees and Hartlepool Port Authority 1989.
15. SECRETARY OF STATE FOR THE ENVIRONMENT. Letter of approval to Cleveland Replacement Structure Plan (see page viii, paras 6.7-6.10), 26 November 1990.
16. JOINT NATURE CONSERVATION COMMITTEE. Directory of the North Sea Coastal Margin. Consultation Draft August 1991.

12. Cardiff Bay - microcosm of conflicts

D. A. CROMPTON, Director of Engineering Operations, Cardiff Bay Development Corporation

Introduction

Cardiff has turned its back on its waterfront. The port has declined with the reduction of the coal and steel industries; large areas of land in south Cardiff are derelict. Cardiff Bay Development Corporation has been established to regenerate south Cardiff. The Corporation's objectives include re-uniting Cardiff with its waterfront. The Development Corporation envisage the creation of a superlative maritime city; the key to that development is Cardiff Bay Barrage. The Barrage will create a fresh water lake of some 200 hectares in south Cardiff and provide a waterfront of 12 kilometres to sustain development.

The Barrage itself has created considerable debate and dissent. There are three main areas of conflict

(a) the Parliamentary process
(b) the need for the Barrage to stimulate development
(c) the environmental issues.

Parliamentary process

A Parliamentary Bill is required to obtain the powers to construct the Barrage. The Parliamentary process has been long. The first Private Bill was deposited in November 1987. This was withdrawn in July 1988 and redeposited in November 1988. That Private Bill completed all but one stage in its process and was talked out by MPs opposed to the Barrage in April 1991.

A Hybrid Bill, promoted by the Welsh Office, is now progressing through Parliament. At all stages of the process, opposition to the Bill has been vigorous. That opposition has taken two forms

(a) opposition by Petitioners whose interests are affected by the Bill; this opposition manifests itself in the Select Committee process where detailed consideration is given not only to the general concept and effect of the Barrage but also to complicated and very technical environmental and engineering issues

DEVELOPMENT PRESSURES

(b) opposition by a limited number of MPs on environmental and financial grounds.

While not wishing to criticise the Private Bill Parliamentary process the change in legislation to reduce the need for Private Bills is to be welcomed. Parliament has obviously considered the issue and takes the view that it is unacceptable that a minority of MPs can successfully defeat a Bill which has majority support in the House.

Economic arguments

One of the major arguments against the Barrage used in Parliament and in general discussion is that the Barrage is not needed to promote development. The argument is that development has started in south Cardiff in advance of the Barrage construction and therefore the Barrage is unnecessary. Significantly the areas where development has commenced are Atlantic Wharf around the closed dock, the proposed new development in the Associated British Ports' site again around the closed dock and the Corporation's development plans with AMEC on a waterfront site. There is clear evidence that development around waterfront is one that investors are looking for at this time.

We had to look very carefully at the economic arguments. Our work has shown that for all the accepted Government economic criteria, net present value, leverage ratio and job creation the Barrage Strategy is superior to both the no Barrage Strategy and to a mini Barrage option put forward by the Royal Society for the Protection of Birds.

The Economic Appraisal has been an invaluable document in refuting opposition claims against the Barrage. The Appraisal is based on a fifteen year development period and takes into account long term growth values rather than addressing the present down turn in the economy. It is perhaps this issue that has attracted the most criticism from opposition.

Environment

It is an environmental issue that has raised the most conflict to the scheme. The environmental oppositon takes three forms

(a) birds and wildlife
(b) water quality
(c) groundwater.

The Taff Ely estuary has been designated a Site of Special Scientific Interest. The designation is based on the use of the Bay for wading birds, redshank and dunlin in particular. The bird numbers approach national importance but in recent years have not reached the level of national importance. The number of birds that currently use the Bay based on average counts amounts to some 5000.

While obviously regretting the loss of the Site of Special Scientific Interest the Corporation has taken the view that the benefit to the redevelopment of south Cardiff outweighs any loss of the feeding grounds. This is an issue that may eventually be debated in the European Court should the Royal Society for the Protection of Birds decide to raise the issue with that body.

Mitigation measures are required to compensate for the loss of the feeding grounds. The Corporation had proposed the creation of an alternative feeding ground some four miles upstream of Cardiff in the estuary. This proposal has recently been rejected by the Select Committee in the House of Commons and further alternative mitigation measures are under discussion between the Welsh Office, the Countryside Council for Wales and the Royal Society for the Protection of Birds.

Water quality has been an emotive issue and led to a great deal of debate. The Taff and Ely rivers are both polluted.

Water quality in the rivers is, however, improving rapidly as work is carried out by Welsh Water and NRA on general water quality enhancement. The Corporation has had to consider the diversion of crude sewage outfalls into the rivers of the Bay within the city of Cardiff itself. We have had to consider the level of oxygenation required in the water to sustain migratory fish which are now using the Taff in particular. We have had to consider the likely growth and methods of dealing with algae within the Bay. We have an adjacent dock, Bute East Dock which is fed with the water from the Taff. None of the water quality problems, the stinking black lagoon scenario, put forward by the Opposition have occurred within Bute East Dock. We are therefore confident that adequate water quality standards can be maintained within the impoundment, and that view is shared by the National Rivers Authority as the regulator.

The third environmental issue which has caused probably the major conflict is the question of groundwater. On impoundment of the Bay the groundwater in Cardiff will rise and it is probable that the rise will have an adverse effect on the basements of some 1000-1500 properties. The technical debates on the amount of groundwater rise have been long and protracted. Virtually every groundwater expert in the country has been involved in that debate. The Corporation, however, are now satisfied that we know what happens with the groundwater in Cardiff; we are confident on the predicted rise and there are engineering measures that can be taken to remedy any adverse effect. There is an extremely comprehensive protection scheme within the Bill involving pre and post impoundment surveys to ascertain adverse effect, a requirement for any adverse effect to be rectified or compensation paid for loss of value together with a wide range of safeguards for property owners in the area.

DEVELOPMENT PRESSURES

Conclusions

While the Cardiff Bay scheme has inevitably brought out large areas of conflict the process needed to obtain powers through Parliament has perhaps exacerbated that conflict. The pressures against the scheme are probably similar to those of any major development but the opposition has been well orchestrated and well managed. The irony of the conflict is that, at the end of the day, the aims and aspirations of the Corporation and the Cardiff Bay scheme are identical to those of the people who are opposing the scheme. Our aim is the long term regeneration of south Cardiff and in that aim there is no conflict.

Discussion

Discussion took place regarding the Towyn disaster and on the responsibility of British Rail for the protection of their track. The need for deep-seated information was emphasised as the cause of the damage was not superficial.

Opinion differed as to the need for additional legislation on coastal planning. One view was that legislation took time and that results were best achieved by talking and working together. This might succeed provided that agreement could be reached on the solution and the financing.

Others held that new legislation was necessary and questioned whether present legislation was used fully and whether planners were using all available information from the coast. There appeared to be no direct contact between Cleveland County Planning Department and the new coastal cell group.

Local Government reorganisation was awaited and might have a major effect on coastal planning.

It was commented that the draft PPG fell short on strategic guidance. Planning authorities in Orkney and Shetland were reported to have jurisdiction up to the 3 mile limit. If applied to the Cardiff scheme this would have produced a local planning enquiry and an examination of the regional impact.

The change in attitude to the conservation of mud-flats at Teesside was encouraging. Conservationists must make an effort to change public attitudes so that mud-flats could be retained. Many problems were due to undervaluing environmental assets and money chasing development jobs often failed. It was noted that property developers were taking a more responsible view of green issues.

13. Coastal planning: recent policy development

J. ZETTER, Assistant Secretary, Department of the Environment

Introduction

The months leading up to the Conference witnessed several major developments in coastal planning policy. The one with which my Division at the Department of the Environment has been associated is the draft Planning Policy Guidance Note on Coastal Planning. This paper concentrates on that particular document. However, it is set in the context of research which was undertaken when ideas for the guidance note were being generated. The paper then goes on to consider recent developments in Europe, which may influence the way we plan coastal areas.

Coastal areas are of particular significance in the United Kingdom. Nowhere in the country is more than 135 km from the sea and the 15,000 km of coastline is much longer, in relation to land area, than for most other countries. The conventional definition of the coast is 'that stretch of land bordering the sea'. But it is accepted that any geographic definition of the coast may include both tracts of land and sea where the terrestrial and marine environments interact. For planning purposes the width of the zone depends on the type of coast and the impact of natural processes and human activities.

The protection of the undeveloped coast is one of the major successes of the comprehensive land use and development planning system introduced 45 years ago. Since that date the sporadic development which characterised an earlier period has been restricted. The planning system has kept the value of land on which it is known that no planning permission will be granted low and this has allowed organisations like the National Trust to acquire and protect a considerable area of coastline.

Local authority boundaries, which derive from the boundaries of Parliamentary constituencies, terminate at the mean low water mark. As Parliament has decided that land use and development planning is, in the first instance, a local matter the effective boundary of the planning system is the mean water mark.

Besides the generality of the planning system, the coast is rich in designated areas to which strict planning policies apply. For example, 5 of the 11 National Parks have coastal stretches, as do 21 of the 39 Areas of Outstanding Natural Beauty. As well as landscape designations, much of the coast is protected by conservation designations. These occur both at the international level by, for example, Special Protection Areas; at the national level by such designations as Sites of Special Scientific Interest and also by local designations included in development plans. In addition there are a number of coastal areas designated as Green Belt.

Even though local planning authority boundaries follow the low water mark, the authorities need to recognise the significant impact that on-shore development can have off-shore. This is particularly important where development proposals span the mean low water mark and need to be taken into account when making planning decisions. Further, for developments inland, local authorities need to have regard to their impacts on coastal areas.

Review of coastal planning

In order to help with the preparation of the draft Planning Policy Guidance Note on Coastal Planning, the Department of the Environment commissioned Rendell Geotechnics to undertake a two part research project. When completed in the summer of 1993 the project will provide a report for planning and management practitioners on the use of earth science information in decision making in the coastal zone.

The Stage 1 report which has now been published reviews the existing situation with regard to the agencies and mechanisms relevant to planning and management in coastal areas. In examining the regulation of development the report looks at development on land, development on the sea-bed, activities on and in the sea, conservation, coastal defence and pollution control. There is also a short section looking at overseas practice, particularly in the United States, Australia and France.

Among the major issues of concern identified in the Stage 1 research report, which are treated in the Planning Policy Guidance Note, are

(a) the need for policies in development plans specifically addressing coastal hazards

(b) the need for co-operation between neighbouring authorities.

However, the major conclusion of the research is that because of the dynamic nature of coastal areas the planning system has to be supplemented by an active management approach.

The Department also commissioned a research report from the UK Centre for Economic and Environmental Development on the major impacts of prospective sea level rise on planning. This identified the planning of coastal areas as a significant issue.

The major implications for land use planning are seen to be

(a) the relocation of infrastructure

(b) the amenity and environmental implications of improved or new sea defences

(c) the need for a pro-active development control policy.

Under a business as usual scenario increased erosion and increased risk of flood will occur and eventually some land and infrastructure will have to be abandoned. If the status quo were maintained then considerable public expense would be involved in improving the level of defence, particularly if some of the higher sea level rise scenarios come about. The report also looks at an enhanced adaptive response. This would involve diverting resources to those areas where the most extensive or valuable assets are at risk, for example, densely populated areas, major communication links and industrial infrastructure. In other areas planned retreat would have to be contemplated.

Planning Policy Guidance Note

Most Planning Policy Guidance issued by the Department deals with specific types of policy rather than with specific types of area. Hence much of the planning advice of relevance to coastal areas is included in other documents.

PPG 1 deals with general planning policies and principles and PPG 2 is concerned with the Green Belt. PPG 7 deals with general countryside matters including National Parks. PPG 12 deals with the general guidance on development plans and development on unstable land is dealt with in PPG 14. Two new PPGs 16 and 17 on respectively archaeology and planning and sport and recreation also have direct relevance to coastal areas.

The draft Planning Policy Guidance Note on Coastal Planning does not repeat other guidance and concentrates on planning policies which particularly arise in coastal areas.

The draft PPG identifies four types of coast which have a relevance for planning policy

(a) the scenic undeveloped coast, protected largely for its landscape value but also for its nature conservation interest

(b) other areas of undeveloped coast, often low-lying areas, which are protected for their high nature conservation value

(c) areas of partly developed coast which may be suitable for development which should be reserved for developments for which a coastal location is essential

(d) the developed coastline, usually urbanised but also containing other major developments, where further development or redevelopment is acceptable or desirable could help improve the environment.

PLANNING AND MANAGEMENT

Three types of policy with particular reference to coastal areas are identified: conservation policies, policies for development and policies for risk.

Conservation policies are, both in the landscape and nature conservation spheres, reflected in the special designations which have been referred to earlier. In many areas these designations overlap. However, some non-scenic areas such as environmentally rich wetlands are either nationally or internationally important as world life habitats. These wetland areas particularly in estuaries are sensitive to proposals for drainage, land reclamation and barrages. Particular care needs to be taken in these areas to assess not just the immediate impact but the wider impacts as well. Also the cumulative effects of a number of small developments is another matter which needs to be carefully considered.

On the development front, coastal areas are subject to a number of physical and conservation constraints. Since the coastal zone is only a part of most local planning authorities' areas, policies to resist development there will be supported if adequate provision has been made in the non-coastal areas of the authority concerned.

Particular types of development identified for treatment in planning policies are tourism and recreation, major developments, mineral development and energy generation.

Turning to policies for risk, the line identified in the draft Planning Policy Guidance Note is to minimise development in areas at risk from flooding, erosion and land instability. Also a decision to protect from erosion or defend from flooding should only be taken where significant development is threatened. Particular areas to which planning policies should pay special attention are low-lying coastal areas, particularly those below the 5 m contour; land close to eroding areas of coastline; and land in coastal areas subject to instability.

The draft Planning Policy Guidance Note also identifies the sort of treatment that coastal policies should receive in the different levels of plan ranging from regional guidance through structure plans to local plans.

Planning the coast is a strategic issue because the scale over which natural processes operate is extensive and often spans regional and local authority boundaries. Coastal planning authorities therefore need to work closely together on coastal plans. The formation of conferences of local planning authorities on a regional and interregional basis, where necessary, is encouraged in the guidance note. Four matters are particularly identified where consistency is needed between neighbouring authorities

(a) developments may bring downstream pollution which damages habitats or recreational resources

(b) development in one authority may reduce the scenic value of coastal areas in another

(c) in the absence of a clear strategy, piecemeal decisions on reclamation of intertidal areas and other developments may damage and erode nature conservation areas
(d) ports, sea defence and recreational developments may alter the natural processes for erosion and deposition or damage areas of nature conservation value.

The research project which was referred to earlier is now looking in detail at the data requirements for coastal planning. However, at this stage information needs can be identified in general terms. The draft Guidance Note deals with these under three headings: physical processes, quality of the coastal environment and development impact. Matters of importance will be recent and current rates of erosion; ecological diversity and other aspects of scientific interest; and the impacts of sewage and waste disposal and in particular of recreation and tourism.

The draft Guidance Note only covers Wales and England. But the former and seven regions in the latter have a coastline and issues of importance will include

(a) the undeveloped coast especially in National Parks and Areas of Outstanding Natural Beauty and Heritage Coasts where landscape conservation is an issue
(b) the treatment of estuaries and estuarine marshes where balancing the needs for special development against the needs of nature conservation and protection of wildlife habitats will often be the main issue
(c) priorities for recreation and tourism and for coast-dependent developments which may have a regional significance, such as ports, oil and gas developments, power stations, barrages and forms of renewable energy generation
(d) coastal defence and the need to take account of the risks of flooding, erosion and land instability where they affect large stretches of the coast.

Structure plans provide the opportunity to set out general policies for the coastal areas of counties. With the help of regional planning guidance and informal conferences of coastal authorities, structure plan policies require co-ordination with neighbouring counties to ensure that a consistent approach is taken. General guidance on planning identifies the need to take an integrated approach to planning policies and this particularly applies in coastal areas.

Local plans are the appropriate level at which to identify a coastal zone to which a specific set of policies may apply. Specific coastal policies will relate to landscape conservation and nature conservation and sites for proposed coast related uses.

PLANNING AND MANAGEMENT

Local planning authorities have a duty to include in development plans proposals for improvement of the physical environment, the management of traffic and for the conservation of natural beauty and amenity. These proposals will be particularly relevant where there is a need to improve and enhance areas of natural beauty, to regenerate run-down coastal towns and ports and to restore stretches of spoiled coastline.

In line with the findings of the research project mentioned earlier the PPG identifies coastal management plans as a way of carrying forward a detailed programme of proposals.

Europe

Almost a year ago the Department of the Environment co-sponsored and assisted at a workshop on coastal zone management held in Poole in Dorset. This brought together delegates from 12 countries including the 11 members of the EC which have coastlines. In his foreword to the report of the workshop Tony Baldry, the Parliamentary Under Secretary of State in the Department, wrote that, 'A great challenge remains in securing the long term health of the coastal zone. There is no doubt that this is a Eurowide if not a worldwide challenge'. It is certainly a field in which the EC has been active.

The Commission has already had significant influence on coastal planning, particularly through the directives on environmental assessment which apply to areas of sea as well as to land, and also through designations, particularly Special Protection Areas.

The Maastricht Treaty mentioned 'town and country planning', in terms for the first time in European law. Some time before that the Commission had published a draft Legal Instrument on the coast but progress has been slow. It is only more recently that the EC has become active again in this area, stimulated by the Dutch who hosted a conference on coastal planning in November 1991 and more recently under the Portuguese Presidency. The Commission plan a directive providing for compulsory strategic plans for sustainable development for the whole of the Community's coasts. Financial support for pilot projects will come from the Financial Instrument for the Environment which has recently been made operational, known more familiarly as the LIFE programme.

With its system of compulsory local plans introduced by the Planning and Compensation Act 1991, the UK is well placed to respond to the Commission's initiative. The position we have taken is that it is preferable to have local plans for whole districts, including the coast, rather than separate plans for coastal areas. There is a need to take a district wide view of conservation and development proposals so that those for the coast can be put in a wider context. The system of structure plans and regional guidance also helps to provide that larger context.

The attitude we have taken on the obligation for these plans to include areas of sea is that there is insufficient experience, anywhere in Europe, on planning areas on the seaward as opposed to the landward side of the coastline. We expect the Commission to stimulate experiment in this area to see what can be learnt about what is a very new area for planning.

Conclusions

Coastal planning policy, like coastal areas themselves, is in a constant process of change. At the time of writing the report from the Environment Committee which is looking at coastal planning and management is awaited and the proposals from the European Commission remain to be fully developed.

In the meantime the draft Planning Policy Guidance Note on Coastal Planning provides for the first time a firm basis for policy. To our knowledge no other country has gone so far as to produce a general statement of this type. We therefore feel there is much to be learned from the approach we have taken. Planning policies and special designations provide significant powers for the protection and enhancement of coastal areas.

But the existing system is kept under constant review. If it is seen to be failing significantly and those concerned can produce good examples at this Conference they will have an attentive audience from those concerned with Government policy in this area. In the meantime it is felt that coastal planning policies and the system of development plans that we have provides a sound basis for pursuing progressive coastal policies.

The views expressed in this paper are not necessarily those held by the Department of the Environment.

14. Coastal defence: legislation and current policy

J. R. PARK, BSc, PhD, FRSC, Head, Flood Defence Division, Ministry of Agriculture, Fisheries and Food

Introduction
This paper concentrates on coastal defence; other matters which are the responsibility of the Ministry such as fisheries and administration of the Food and Environment Protection Act will be referred to where they are relevant to coastal defence. Legislation relating to flood and coastal defence was thoroughly reviewed in a paper presented to the ICE Coastal Management Conference in May 1989 (ref. 1). At the time of preparation the Water Bill was before Parliament. Subsequently the Water Act 1989 was passed by Parliament and the National Rivers Authority was established with a duty to oversee river management and regulation, and with strong environmental responsibility. Flood including sea defence became the responsibility of the NRA and as with the Water Authorities the function is discharged through regional and local committees, now called Flood Defence Committees. In 1990 the Law Commission embarked on a consolidation of legislation relating to water. Flood defence legislation, in the Land Drainage Act 1976 and the Water Act 1989, was re-enacted in the Land Drainage Act 1991 and the Water Resources Act 1991.

Organisation
A short summary is presented to provide the context. The Ministry of Agriculture, Fisheries and Food has policy responsibility in England for flood defence (defined as the prevention of flooding of land by rivers or the sea) and, since 1985, coast protection (defined as the protection of land from erosion or encroachment by the sea). In Wales the Secretary of State for Wales holds the equivalent responsibility.

Flood defence legislation provides for the construction, improvement and maintenance of defences against inland and coastal flooding in England (and Wales) to be undertaken by the National Rivers Authority through its ten Regional Flood Defence Committees (RFDC) whose Chairmen are ministerial appointees; by local authorities; and by some 250 Internal Drainage Boards (IDBs) which administer areas with special drainage needs. Coast

PLANNING AND MANAGEMENT

protection legislation empowers maritime district councils to carry out works to protect against coastal erosion. Under relevant Acts, it is generally the responsibility of the appropriate authority to identify the need for works and to decide which projects it wishes to promote.

These arrangements give emphasis to local knowledge. The RFDCs comprise representatives of the area covered and include members appointed by constituent councils, the NRA and the Minister. Since the committees were re-constituted under the Water Act 1989, the members appointed by the Minister to each committee include a representative of conservation/environmental interests nominated by English Nature and the Countryside Commission in consultation with local bodies. Representation on IDBs was changed when the new funding arrangements were introduced in 1990. IDBs now comprise elected members drawn in the main from drainage ratepayers and also members appointed by local authorities in proportion to their special levy contributions to the funding of the boards' expenditure (currently up to 40% of the full board but up to a single majority from April 1993).

The NRA has a duty to exercise a general supervision over all matters relating to flood defence and to carry out surveys. In addition, coast protection authorities are required to consult NRA and neighbouring authorities. A recent development encouraged by the Ministry is the establishment of Coastal Groups with representatives of coastal defence authorities (including local authorities and the NRA) which encourage co-ordination, exchange of information and discussion of issues of common concern; the majority of the coastline of England and Wales is now covered by such groups. MAFF and the Welsh Office have established a Coastal Defence Forum involving representatives of coastal groups and the NRA with the aim of enhancing co-ordination and providing a lead on strategic planning of coastal defences on a coastline basis.

Environmental considerations

Flood and coastal defence works, both capital schemes and maintenance, can have a major impact an flora, fauna, historic and archeological remains and the landscape unless they are carried out with care and understanding. The booklet entitled "Conservation guidelines for drainage authorities", prepared jointly by the Ministry, the Department of the Environment and the Welsh Office, provides guidance on fulfilling statutory duties. It sets out what should be done by drainage authorities rather than how it should be done and stresses the need for early consultations with environmental bodies about proposed works, both capital and maintenance.

A revised version was produced in December 1997, following consultation with conservation bodies and operating authorities. The draft does not deal specifically with coastal protection works which are the subject of separate legislative requirements but local authorities are strongly advised to follow the guidelines when carrying out such works as well as flood defence operations.

One of the changes since the booklet was last published in 1988 is that regulations have been made implementing council directive 85/337/EEC on the assessment of the effects of certain public and private projects on the environment. New flood defence and coast protection works require planning permission. Specified projects are subject to the Town and Country Planning (Assessment of Environmental Effects) Regulations 1988, and it is for the planning authority to ensure that environmental factors are fully considered for other projects. Flood defence improvement works which have deemed planning permission are subject to the Land Drainage Improvement Works (Assessment of Environmental Effects) Regulations 1988. In both cases the regulations provide for the advertising of proposals, consideration of representations, preparation of environmental statements and reference to Ministers for decision in cases of dispute.

Grant aid

The Ministry is empowered to pay grant in respect of flood and coastal defence works. NRA and local authorities are required to prepare forward plans. Officials monitor and evaluate those plans in order to estimate the level of Government support necessary to take priority works forward, to approve schemes and to allocate grant. Government funding for flood and coastal defence in England has been increased on five occasions since 1986/87. The increases arise from assessments of the current state of flood defences which have demonstrated the need for significant additional capital works in order to reduce the risk of flooding. The extra grant for coast protection is primarily for the repair of defences damaged in the winter storms of 1989/90.

Priority for grant is given to flood warning systems and capital works designed to protect people and property. Thus urban rather than rural areas are the main beneficiaries. Sea defence schemes have priority over flood protection. Lower priority is given to rural sea defence and the continuance of existing rural drainage schemes, with new rural drainage schemes having the lowest priority.

In order to qualify for grant, schemes are required to be technically sound, economically viable and environmentally sympathetic. The Ministry's Regional Engineers liaise closely with authorities and are able to offer advice in the early stages of scheme preparation. Major changes have occurred during the past decade in the way in which coastal defences are engineered. There has been a swing away from the construction of sea walls towards more natural forms of defence such as a beach which has the ability to absorb

PLANNING AND MANAGEMENT

Table 1*

Sea defence and tidal works		Coast protection
Districts of NRA %	LAs %	%
35		35
45	45	45
55		55
65		65
75		75
85		

* Internal Drainage Boards would attract 50% grant for sea defence and tidal works.

energy in storms and to rehabilitate itself. Grant aid is available for beach renewal. Grant aid is also available towards the funding of preliminary investigations, including feasibility and strategic studies such as the East Coast Sea Defence Management Study carried out on behalf of the Anglian Region of the NRA. No scheme is approved for grant aid unless English Nature have indicated that they have no objections to the proposals.

Value for money is an important consideration in the investment of taxpayers' money. However, emphasis is given to ensuring that all benefits are identified. For instance at Aldeburgh where the scheme would not have been justified on urban and agricultural benefits alone, environmental and recreational benefits were included in the assessment. Quantification of such benefits is at the limits of current methodology so the Ministry is funding relevant research. Much of that research has been carried out at Middlesex Polytechnic and they are in the process of preparing a report of the research and drawing together the results in the form of a manual for use by authorities.

In the 1991 Public Expenditure Survey, the Government renewed its commitment to the programme of replacement of ageing defences and extended the programme to 1994/95. To assist drainage authorities in completing urgent works, increased grant rates will be introduced in 1992/93. In particular, a new grant rate of 85% (including the 20% supplement for sea and tidal work) will be created for schemes undertaken by the National Rivers Authority in Lincolnshire - an area where the programme of work is set to nearly double between 1991/92 and 1994/95. Also a new grant rate of 45% (again including the supplement) has been created to assist the National Rivers Authority in some other districts. The opportunity is also being taken to rationalise the grant rates available for flood and coastal defence works carried out by other authorities. The full range of grant rates available for sea defence and tidal works (including the supplement) and coast protection is as shown in Table 1.

In addition to the two new NRA rates mentioned above, the other changes from the rates of grant available currently are

(*a*) coast protection - current range of 15 grant rates rationalised to 5 grant rates

(*b*) local authorities and Internal Drainage Boards - rates reduced by 1%.

The rate of grant applicable to individual districts of the NRA or to individual coast protection works is determined by a formula relating the cost of the works to the relevant community charge population. The aim is to provide the highest grant support where the needs are high and local resources are low.

Taking into account the outcome of last year's Public Expenditure Survey, grant provision is

	1991/92	1994/95
NRA flood and sea defence	30.1	48.2
Local Authority flood defence and coast protection	22.9	22.7

Response to climate change

It is clear that climate change will have an effect on mean sea level and weather patterns, although the detailed effects remain unclear. A rise in sea level globally will obviously have an impact on coastal areas in this country and predicted changes in climate may also affect the frequency and intensity of storms.

The Ministry strategy for tackling sea level rise promulgated in July 1989 and endorsed by the House of Lords Select Committee on Science and Technology is to

(*a*) refurbish defences to reduce existing risk

(*b*) continue research into river and coastal processes

(*c*) monitor trends in climate, sea level, waves, beaches and saline intrusion

(*d*) utilise current predictions of sea level in a review of existing standards

(*e*) keep policy and best practice under review as understanding develops of sea level trends, surges and weather patterns.

The strategy also encourages designs which allow for adjustments when predictions are clearer.

PLANNING AND MANAGEMENT

The Inter-governmental Panel on Climate Change (IPCC) have since reported offering predictions of global sea level rise for various scenarios. The "IPCC best estimate trend" which suggests a rise of about 20 cm by 2030 and 65 cm by 2100 is accepted as the most appropriate at this time. These predictions have been utilised with estimates of earth crustal movements in Great Britain in a review of existing standards to provide the best basis of allowances for the design of coastal defences. The allowances based on NRA Regions are

Anglian, Thames, Southern	6 mm per year
North West, Northumbria	4 mm per year
Remainder	5 mm per year

This advice which was prepared in consultation with NRA was issued to all drainage authorities in November 1991.

Research and development

Design decisions regarding the effective provision of flood and coastal defence need to be based on an up to date understanding of natural processes and engineering options. The Government supports a wide ranging R&D programme, administered by the Ministry, of some 2.5 m a year. The programme is guided by a medium term strategy and particular emphasis is given to studies of river and coastal processes, of the environmental impacts and the implications of global warming. The Ministry draws on the advice of an expert committee; this advice has recently been updated. Co-ordination arrangements between the major sponsors are in place in order to avoid duplication and to strengthen the important links between basic science and applied research.

Current actions

Interest in coastal zone management is increasing substantially and coastal defence is one of a number of activities that affect coastal zone management. Currently this topic is the subject of a House of Commons Environment Select Committee Study. The Committee have taken written evidence from a number of parties and are taking evidence from witnesses. The study is likely to lead to a report in Spring 1992. The full range of interests is inputting to the study. The report will provide an independent assessment of the area. Government will be required to respond in due course. The process is unlikely to be complete by the time of the Conference.

The National Audit Office is carrying out a value for money study of coastal defences, including sea defence and coast protection. This has been under way since Autumn 1990 and is moving towards a report. The report is expected to focus on the effectiveness of policy implementation for coastal defences. The study is restricted to coastal defence.

Current issues

The foregoing has identified a number of issues which are already evident and are expected to continue to be of interest. The current framework provides a continuing working basis for taking forward action related to those issues. Thus environmental considerations currently carry significant weight in decisions relating to flood and coastal defence policy. Their nature is likely to be better understood and hence appropriate responses may change in emphasis, but the current system is capable of responding accordingly.

Climate change will clearly continue to influence decisions on policy. The framework established through the Ministry strategy policy provides a basis for decisions in the future. IPCC work will be refined such that predictions will be possible for sea level rise on a regional basis and suggestions of changes in weather patterns will be clearer. This development of understanding can be utilised in updating the strategy.

R&D will continue to provide a basis for judging the appropriate response. This will clearly influence the assessment of the necessary programme of works to defend the coastline. R&D will also identify possible new approaches to defence and will increase the effectiveness of designs. As the use of soft engineering solutions gains priority this will focus attention on the appropriate materials and their sources, including the sensitive question of dredging for marine aggregates.

A particular emphasis is likely to be placed on the strategic approach to planning of policy and implementation. Time is being taken to identify those issues which are of a strategic nature. These will influence the development of policy. It is already clear that strategic planning will influence the choice of technical options for coastal defence, in the context of the coastline as a whole. The creation of Coastal Groups and the Ministry/Welsh Office Coastal Defence Forum provides a mechanism for developing this approach. A wider range of interests and a potentially longer list of technical options are likely to be a feature of coastal defence planning in the future.

One very important linkage is covered by the juxtaposition of this paper and that from the DoE. There is already an important inter-dependency between coastal defence planning and development control. Current guidance is under review and I expect the guidance on development in areas of flood and erosion risk to be clarified, not least arising from the potential impacts of climate change.

It is obvious that change is in the air as far as coastal defences are concerned. Much has happened in recent years. Some of the future influences

PLANNING AND MANAGEMENT

have been identified. Much of this pressure will be along directions which have already been identified. The newest feature is the emphasis on coastal zone management and for coastal defences to be considered in that context.

(The views expressed are those of the author and not necessarily those of the Ministry of Agriculture, Fisheries and Food.)

Reference
1. Institution of Civil Engineers. Coastal management. Thomas Telford, London, 1989.

15. Overseas responses through legislation and policy to coastal problems

J. C. DOORNKAMP, MSc,PhD,CGeol, Department of Geography, University of Nottingham and Consultant to Rendel Geotechnics

Introduction
In any consideration of planning and management of our coasts it is worth standing back for a moment in order to examine some of the trends in the administrative responses made by other countries, many of which face similar problems to our own.

Inevitably this review must be brief and selective. It is neither definitive nor comprehensive. However, it does indicate some important trends in the administration of the coast in other countries.

Most of the examples are taken from the USA, Australia, and a number of European countries.

Rather than providing a country by country description this review looks at legislative and policy responses to particular coastal problems. In particular, it examines responses to the problems of

(a) maintaining navigable waterways and harbours
(b) degrading coastlines
(c) floods and flood control
(d) pollution
(e) conservation of valued ecosystems
(f) potential problems in the event of a rising sea level
(g) adopting a holistic approach to coastal management
(h) funding coastal works
(i) devising an appropriate administrative structure
(j) the EC context.

Throughout this account it has to be borne in mind that legislation and coastal management policy have to be seen in the cultural context of the society within which they are formulated. Some of the problems are not so much a problem of the coastline itself but of the administrative framework within which policy is being administered. For example, it we want to make comparisons between the administrative system in Britain and some other country we might not choose another European country (e.g. France) where the structure of the administration is rather different, but the USA where the *Federal - State - Township/City* hierarchy of administration does have some close parallels with our own hierarchical structure of *National - County - District*.

The other factor to bear in mind throughout these comparisons is that there is a considerable difference in the history of coastal land uses around the world. This has undoubtedly affected the nature of some of the administrative responses.

Legislative responses to particular problems
Maintaining navigable waterways and harbours

One of the oldest statutory instruments concerned with the coast is the Rivers and Harbours Appropriation Act (1899) in the USA which empowered the US Corps of Engineers to issue or refuse permits for the construction or alteration of all built structures within or adjoining navigable waterways. It also precluded the alteration of channels unless duly authorized. This has now been extended to include similar authority along shorelines, and within wetlands and dunes.

This very early Act (1899) (though Norway had introduced an Act on Sea Transportation some 6 years earlier) anticipated the need to manage the dynamic aspects of the coastal system.

Degrading coastlines

Degrading coastlines can take a number of forms (e.g. rapidly retreating cliffs, erosion of spits and bars, breaching of sand dunes), and in many cases pose a threat to the established land uses and human occupance of the area. There have been a number of reactions to such situations, ranging from the recognition that such events must be resisted (usually by providing some form of protection) to the idea that they must be allowed to take their natural course.

In the USA, in such situations there has been a tendency to adopt coastal zoning strategies whereby a "coastal construction setback line" is defined. Such a zone not only provides a buffer between new buildings and an active coastal system (e.g. one where cliff retreat could cause property damage within the economic life of the building) but it also prevents construction from having a damaging effect on the coastal system itself (e.g. through increased ecological pressure on dune systems rendering them unstable).

Similarly in South Australia, in areas of coastal erosion setbacks are adopted and it is specifically stated that they must allow for an anticipated 100 years of erosion.

In Queensland, Australia, the Beach Protection Act of 1968 established both the Beach Protection Authority and the Beach Protection Advisory Board, each with its own powers and functions to deal with coastal degradation. The primary purpose of the Authority is to provide for the protection of beaches against, and for the restoration of beaches from, erosion or encroachment by the sea. The purpose of the Advisory Board is to provide guidance and advice to the Authority.

The functions of the Authority are detailed within the Act to include

(a) the carrying out of investigations with respect to erosion or encroachment by the sea
(b) the investigation and planning of preventive and remedial measures
(c) the recording and evaluation of the results of such investigations and plans.

The Board, on the other hand, is concerned with

(a) considering any matters pertinent to the protection, from erosion, of beaches generally or any beach in particular
(b) the definition of beach erosion control districts
(c) making recommendations to the Authority in respect of either of these two matters.

Crucial to the operation of the Act is the definition of Beach Erosion Control Districts, for it is with these that the Authority has special concern. The matter is left in the air by the Act in that the Authority may recommend to the Governor at any time that any part of the coast should be so designated.

Once the District has been declared the Authority is required to prepare a scheme for the protection of the beaches against erosion or encroachment by the sea. There is a right of complaint against such plans by anyone who feels aggrieved. However, there is no public enquiry, only a receipt of the complaint in writing and a discussion of the complaint by the Authority, which may or may not then alter its plans.

In Denmark in 1981 a Government Circular was issued which established a protection zone along the coast that prevented the building of holiday and recreational homes within 3 km of the coast. This followed the limitations on holiday home development imposed by the Urban and Land Zone Act of 1970, and prohibitions issued by the Ministry of the Environment in 1977 (prohibited zoning of new summer holiday cottages) and 1978 (prohibited the construction of new holiday hotels along the coast). Issues other than the protection of degrading coastlines were also involved.

Floods and flood control

Coastal dynamic processes in many parts of the world have created natural barriers against flooding in the form of storm beaches and sand dune

PLANNING AND MANAGEMENT

systems. It is clearly in the interests of coastal management to not only preserve such natural flood defences but also to allow them to go on regenerating and growing wherever they are part of the present-day suite of coastal processes.

Some countries have found it necessary to legislate to bring the point home to coastal users. For example, in the USA the Coastal Barrier Resources Act (1982) draws particular attention to the management of natural protective barriers to flooding and coastal erosion.

Pollution

Legislation to deal with water quality and the disposal of wastes through discharges that affect rivers, estuaries and seas have become common place around the world, e.g. the Federal Water Pollution Act of 1972 in the USA; and in Norway both the Act on Protection Against Oil Emissions Damage (1970) and the Pollution Control Act (1981).

Conservation of valued ecosystems

There is a general recognition in many parts of the world that the human use and exploitation of the coastal zone has led to increasing conflicts between alternative uses, and has caused ecological and environmental damage along the way. The fragility of much of the coastal system, including its marine life, is recognised

> "Important ecological, cultural, historic, and aesthetic values in the coastal zone **which are essential to the well-being of all citizens** are being irretrievably damaged or lost" (my emphasis).

> "Special natural and scenic characteristics are being damaged by ill-planned development that threatens these values."

Part of the blame for this state of affairs is assigned to the inadequacy of existing legislation at both the state and local level.

The answer is seen to be in a more direct control, by the competent administrative authorities, of developments within the coastal zone

> "The key to more effective protection and use of the land and water resources of the coastal zone is to encourage the states to exercise their full authority over the lands and waters in the coastal zone by assisting the states, in cooperation with Federal and local governments and other vitally affected interests, in developing land and water use programs for the coastal zone, including unified policies, criteria, standards, methods, and processes for dealing with land and water use decisions of more than local significance."

(1988 Edition of the *United States Code*, Government Printing Office, Washington, Chapter 33).

In other words, there is a recognition that Government at all levels should co-operate in a unified approach to coastal management. This approach is to include a unified set of goals and a unified approach to dealing with particular situations. However, there still remains the problem of deciding what is of "more than local significance".

The declaration of policy by Congress goes some of the way towards defining those areas which have a wider significance.

"The Congress finds and declares that it is the national policy - ...(2) to encourage and assist the states to exercise effectively their responsibilities in the coastal zone through the development and implementation of management programs ...which programs should at least provide for -

(A) the protection of natural resources including wetlands, flood plains, estuaries, beaches, dunes, ... and fish and wildlife and their habitat within the coastal zone.

(B) the management of coastal development to minimise the loss of life and property caused by improper development in flood-prone, storm surge, geological hazard, and erosion-prone areas and in areas of subsidence and saltwater intrusion, and by the destruction of natural protective features such as beaches, dunes, wetlands and barrier islands."

The Congressional view is that the administration of the coastal zone within these policies and declarations, as well as the Act, should not be adversarial but through consultation

"(4) to encourage the participation and cooperation of the public, state and local governments, and interstate and other regional agencies, as well as the Federal agencies having programs affecting the coastal zone, in carrying out the purpose of this chapter."

There have been specific congressional initiatives in respect of ecologically valuable coastal areas. For example, special studies have been initiated such as that on estuaries, in order to assess their specific characteristics and value to the nation (Public Law 90 - 454 (1968) see *United States Code*, 1988 Edition, Vol 6, Page 1187). Others include studies of important and large areas of special coastlines such as Chesapeake Bay - "the greatest natural ecological entity of its kind in the United States" - for which a co-ordinated research programme was set up at federal level through the Chesapeake Bay Research Coordination Act of 1980.

Another example relates to natural coastal barriers, mentioned above in the context of flood protection. In a sense the Coastal Barrier Resources Act

PLANNING AND MANAGEMENT

(1982) in the USA which paved the way for some of the thinking, became incorporated in the Coastal Zone Management Reauthorization Act of 1985. Nevertheless, coastal barriers came in for special treatment under the 1982 Act, especially in the context of restricting any actions that affected this "Coastal Barrier Resource System" and in particular bringing about the long-term conservation of such areas.

By this means a single, physically defined, unit of the coastal system was singled out for special protection and placed in a special legal category. The function of coastal barriers was seen not only in ecological terms but also in terms of their role in coast protection

> "coastal barriers serve as natural storm protective buffers and are generally unsuitable for developments because they are vulnerable to hurricane and other storm damage and because natural shoreline recession and the movement of unstable sediments undermine manmade structures."

It was recognised that earlier permitted developments had been foolish, and had cost millions of tax dollars in that many had involved federal subsidies. The purpose of this Act was to turn the government away from subsidising adverse influences on a sensitive and dynamic part of the coastal system.

The Swedish Environmental Protection Agency (SNV) has made an inventory of valuable marine and coastal areas that have been little affected by human interference. On 1 September 1990 a *National Plan for Nature Protection* was published by SNV, and the implementation of this protection is made possible by four separate Acts: the Natural Resources Act, the Planning and Building Act, the Nature Conservancy Act and the Environmental Protection Act.

It is also recognised, that to be effective, protection has to be monitored through an inspection programme, and this has been built in to a Management Plan. By early 1991 some 80 sites or so had been listed for protection.

Choice of sites was made within the context of the variety of coastal types within Sweden (based largely on their geomorphological and biological characteristics), using the IUCN criteria for the planning and management of marine and coastal protected areas, viz

(*a*) selection criteria: genetic pool, naturalness, diversity, extent, natural unit, integrity, rarity, uniqueness, representivity, survival, criticalness, dependency, scientific value

(*b*) priority selection: threat, fragility, information value, educational value, recreational value.

Protection has a knock-on effect on the human activity that affects the system within which the protected area lies. For example, protection requires the better management of waste water that can enter the system from outside the protected area but can, nevertheless, affect the protected area as currents carry the waste along.

Relevant legislation in other countries includes the Nature Conservation Act (1970) in Norway.

In the Netherlands, of course, there is a very practical reason for wanting to protect the ecologically-valued dune systems of the coast. These dunes are a crucial part of the natural defenses against flooding from the sea. For example, the central government in The Netherlands has assigned the dune area as one which serves a primary national function, and therefore no urban encroachment onto the dunes is permitted. Nevertheless, there is still a fragmentation in terms of management responsibility (e.g. in respect of the all important coastal dune system) together with an insufficient legal protection of the natural value of the dunes. The latter is evident in the recreational and commercial developments that have managed to become established within this coastal zone.

This has been corrected, in part at least, by *The Nature Conservation Policy Plan* (1990) which contains very specific proposals about actons that are permitted in the dune areas. This will be brought about by the specific and systematic application of the Nature Conservation Act to the dunes. This will make illegal any activities that are harmful to nature and the dune environment.

In France under the 1986 Law (see below) town and country plans must address environmental considerations. A plan for a coastal area must include the concept of a "threshold capacity" - through which there is the mechanism to protect environmentally and ecologically sensitive areas, to preserve agricultural land and to provide adequate public facilities. This relates to set-back provisions and the prohibition of road building within 2 kms of the shore, and of any structure on beaches, dunes, lagoons or any other part of the coastline.

Potential problems in the event of a rising sea level

In the event that a rising sea level may bring significant changes to the coast a number of countries have recently sought to adopt coastal management strategies that are likely to be appropriate.

In the case of South Australia, for example, coastal planning constraints have been adopted which are equivalent to the setback zoning approach adopted in parts of the USA. For example, in South Australia a development must be above the mean high-water level plus the height of a 1-in-100-year storm surge, plus another 300 mm to accommodate a greenhouse effect rise, and a 250 mm safety margin. In addition there must be space in front to construct a bund for protection against a further 1 m rise. This bund, however, must not be deleterious to public amenity and must not affect any beach or damage the coastline.

PLANNING AND MANAGEMENT

Adopting a holistic approach to coastal management

The days when hard engineering was the reflex action to a coastal problem have long since gone. It went with the realisation that the complex facets of the coastal environment and the great variety of human actions within and desires for coastal areas requires a holistic approach to coastal management.

In the USA the National Environmental Policy Act (1969) was designed to

> "encourage productive and enjoyable harmony between man and his environment: to promote efforts, which will prevent or eliminate damage to the environment ... to enrich the understanding of the ecological systems and natural resources important to the Nation, and to establish a Council of Environmental Quality."

Ever since then the holistic view has tended to prevail within coastal management in the USA.

For example, the Coastal Zone Management Act of 1972 (Public Law 92-583) in the USA (which was amended in 1974, 1976 and 1978, and received both amendments and extensions in 1980 and 1985) was one of a number of legislative instruments in the USA which brought about a more comprehensive management of the coastal zone.

In Australia similar attitudes are also in evidence. In 1990 the New South Wales Government released its Coastal Policy, the stated aim of which is to

> "Protect the coastline and beaches for the enjoyment of future generations and to ensure that coastal development is balanced, well-planned and environmentally sensitive."

The principle elements of the policy are to

> (*a*) continue the existing State programme to bring unique coastal land into public ownership
> (*b*) limit urban coastal development to areas adjacent to existing towns
> (*c*) cluster tourist development
> (*d*) make the height and concentration of developments sensitive and appropriate to the local environment
> (*e*) protect representative coastal species and ecosystems through the continuation of existing wetland and littoral policies
> (*f*) develop a coastal hazards policy to assist local governments in having to deal with natural coastal hazards and processes.

In Western Australia it has also been declared that the coastline is a matter of national (rather than state) concern. This was expressed in 1991 in the House of Representatives Standing Committee on Environment, Recreation and the Arts report: *The Injured Coastline*. This contains a strong argument in favour of an holistic approach to the coastline, replacing the existing piecemeal (and they conclude largely ineffective) approach. There is an accusation that up till now coastal management has dealt with resolving specific problems rather than with the management of the complex whole, such management requiring an improved co-ordination at all levels of government and between many disciplines.

There is a recognition that problems are compounded by the fragmented nature of the decision-making process, the multiplicity of public agencies, the existence of arbitrary administrative boundaries and the failure to consider the cumulative effects of individual decisions.

The Committee proposed that

"The Commonwealth develop without further delay a national coastal zone management strategy in co-operation with the States and Territories and local governments to provide a framework for the co-ordination of coastal management throughout Australia. The strategy should incorporate agreed national objectives, goals, priorities, implementation and funding programs and performance criteria."

The responsibility for developing a national coastal strategy should lie, according to the Committee, with the (existing but newly constituted) National Working Group on Coastal Management.

In Denmark in 1990 the Ministry for the Environment prepared a Circular the purpose of which was to ensure that seaside development is environmentally sound and takes due consideration of natural resources, so that the coasts continue to be naturally balanced and attractive for recreational purposes - while also remaining spacious enough for future business functions that depend on a coastal location. In other words, once again there is a national recognition of the need for a holistic management approach to a multi-faceted environment.

Funding coastal works

The Coastal Zone Management Reauthorization Act of 1985 extended the 1972 Coastal Zone Management Act for a further 5 years and provided new guidelines concerning the conditions under which Federal funds would be provided as a contribution towards State expenditure on coastal works.

Under the 1985 Act grants (up to 80% of a defined set of costs in any one year) can be made by the Federal Government to assist coastal states in their administration of the coastline. However, in order to qualify a state has to prepare a management programme which includes

(*a*) an identification of the boundaries of the coastal zone subject to the management programme
(*b*) a definition of what shall constitute permissible land uses and water uses within the coastal zone

(c) an inventory and designation of areas of particular concern within the coastal zone

(d) an identification of the means by which the state proposes to exert control over the land uses and water uses ... including a listing of relevant constitutional provisions, laws, regulations, and judicial decisions

(e) broad guidelines on priorities of uses in particular areas, including specifically those uses of lowest priority

(f) a description of the organizational structure proposed to implement the management programme

(g) a definition of the term "beach" and a planning process for the protection of, and access to, public beaches and other public coastal areas of environmental, recreational, historical, aesthetic, ecological, or cultural value

(h) a planning process for energy facilities likely to be located in, or which may significantly affect, the coastal zone, including, but not limited to, a process for anticipating and managing the impacts from such facilities

(i) a planning process for assessing the effects of shoreline erosion (however caused), and for studying and evaluating ways to control, or lessen the impact of, such erosion, and to restore areas adversely affected by such erosion.

This last paragraph is interesting in that it implies an acceptance of a philosophy of coastal preservation rather than one of allowing coastal evolution. It also implies the notion that erosion should be judged in terms of its site and not its situation. The emphasis here is on the area affected by the erosion rather than on the context of that event within the coastal system as a whole. (In this author's view scientific research and engineering experience would suggest that such an assumption is unsupportable.)

An important effect of the rules governing the allocation of grants is that non-engineering remedial strategies are emphasised and have to be considered before an engineering reaction can be contemplated.

In Queensland the Beach Protection Act of 1968 enabled funding to be obtained through the Beach Protection Fund. Sources of money for this Fund include loans (mainly from the Treasury), money appropriated by Parliament and moneys paid to the Authority. Some of this money may be awarded as grants to a Local Authority carrying out approved coast protection works.

In France the Coastland Conservation Institute has the power to purchase land so that undue pressures by developments on the coast could be prevented. In particular, it has the power to purchase areas of unquestionable ecological interest.

The Conservancy makes these purchases on the basis of advice from the municipal council of the commune concerned, and has to resort to purchase through expropriation, rather than through amicable agreement, in only 6% of all cases.

The Conservancy is organised into seven regions, and it has both administrative and financial autonomy. Most of the funds (about 100 million FF) are derived from the State. It is headed by a Board of Directors consisting of 34 members (half coming from coastal "regions" and "departments", the other half representing special interests in the coastline).

By the end of 1990 the Institute had acquired 36,000 hectares of land at 276 sites, which represents 473 linear kilometres of coastline. The land purchased is managed by the commune(s) concerned or by a management association.

Devising an appropriate administrative structure

Much has been made here of the position established in the USA by the Coastal Zone Management Act of 1972 and by the Coastal Zone Management Reauthorization Act of 1985. In terms of administrative structures and responsibilities the underlying principle in these Acts is that the State has local control but must stay within national policy concerning the coastal zone. The federal government in its turn has to ensure that federal activities directly affecting the coastal zone are consistent with state programmes, and will avoid inconsistencies with state coastal zone management policies.

It is important to note the responsibilities which Congress declared that a state must accept. These include

(a) protection of fish, wildlife, and natural wetland resources

(b) minimizing the loss of life and property from coastal hazards

(c) establishing guidelines for siting major energy, fisheries, recreation, ports, and transportation facilities

(d) assuring that local regulations do not restrict public access and recreation

(e) redevelopment of urban waterfronts and ports

(f) preservation and restoration of historic, cultural, and aesthetic coastal features

(g) consulting and co-ordinating actions with federal agencies

(h) encouraging public and local government participation in coastal management decision making

(i) establishing comprehensive conservation and management plans for living marine resources, and siting pollution control and aquaculture facilities (U.S. 1982: s.1452 (2)(A)-(I).

In respect of the Queensland beach erosion control legislation (see above) there are times when the beach erosion control district extends across the boundary between two Local Authorities. In such cases the plan is imposed on the Local Authorities with each required to carry out any necessary works within its own administrative area.

PLANNING AND MANAGEMENT

However, one of the critical issues is the extent to which other countries have adopted national rather than local control over management of the coast. For example, in France, where there has been a political trend towards decentralisation of government, the primary responsibility for the coastline has been retained by Central Government. Awareness of the need for rational development of the coastline became evident in the early 1970s following a 1972 report which projected the long-term future of the coast of France. In 1977 the President declared that coastal areas were a matter of national concern, a matter that was brought into legislation through the "1986 Law" (i.e. the Management, Protection and Development of the Seashore Law passed in January 1986).

Coastal management in France includes the

(a) management by Central Government of nationally-owned coastal property
(b) establishment of stricter planning regulations along the coastline
(c) purchasing of countryside along the coast by the Institute for Conservation of the Coastline and Lakesides (CELRL - Le Conservatoire de l'Espace Littoral) - a public body established by law on 10 July 1975
(d) planning conflicting coastal uses through Marine Development Plans (SMVS).

In 1982/83 planning was devolved to the individual communes which became responsible for their own Land Occupation Plans (POS) and for the issue of building permits. The 987 coastal parishes have been compelled to prepare their own POS (Plan d'Occupation des Sols), which previously had been required only for large cities, and in the case of coastlands they had to be prepared in accordance with the Schema d'Aptitude et d'Utilisation de la Mer (SAUM) where one exists.

However, the "coastal law" does set the framework for local action. This had already been set down in a Directive (4 August 1976) which called for

(a) the inclusion in plans of open spaces which contrast with any proposed development
(b) a limitation on development close to the shore, and a justification for any such developments which are allowed
(c) the retention of free access by the public to the shore through any new development
(d) the prohibition of any new developments within 100 metres of the high water mark
(e) the inclusion within coastal development plans of areas reserved for camping and caravanning
(f) the preservation of ecologically or scenically special areas
(g) a prohibition on road building within 2 km of the coast.

In addition the "coastal law" increased the provisions for public enquiry and the opportunity for public participation in debates about developments in the coastal zone.

It is clear that in France it is now recognised that appropriate coastal management areas may not coincide with long-established administrative boundaries.

The EC context

Our position in Europe is no longer just one of sharing a common continental shelf with several of our neighbours, and thereby sharing large marine dynamic systems: it now includes the regulatory controls emanating from the Parliament of the European Community.

Thus, while individual European countries have brought in their own policies and legislation in respect of coastal management, their individual concerns can now find a wider expression through the Directives of the European Parliament. It is important to realise, therefore, that the (usually bad) experiences of some countries may affect the final nature of some of the EC pronouncements. For example, the European dimension shows that the coastal zone cannot be considered in isolation from the mainland, and in particular cannot be considered in isolation from major influences such as large river systems that supply sediment to the coastal system. Unfortunately these rivers also supply waste and pollutants that can have a damaging effect on water quality and sensitive environments within the coastal zone. For example, a series of chemical spillages along the Rhine in 1986, 1987, and 1988 affected water quality not only in the Rhine delta and the immediately adjacent shoreline: its influence reached the Wadden Sea, decreasing water quality and placing the ecological community at risk. It is not surprising, therefore, that the EC is concerned about pollution, ecological protection, and the impacts that events in one country can have on another.

There appears to be a growing recognition within the EC that cross-border matters in the environmental field should be dealt with by legislation at Community level. This was so in the case of the introduction of Environmental (Impact) Assessments in such matters as air pollution. It now seems to be a principle to be adopted in terms of the marine environment.

To some extent this type of cross-national approach has its earlier examples in agreements between the Nordic Countries, for example in relation to marine pollution. Denmark, Germany and The Netherlands have a co-ordinated policy on the management of the Wadden Sea.

The Community has already published Community-wide surveys of such elements of the coastal system as water pollution and the cleanliness of beaches. Further moves that will affect coastal management in the UK, as well as elsewhere in Europe, are discussed below.

PLANNING AND MANAGEMENT

It is important to recognise that there are a number of pressure groups at work within Europe that have a considerable influence on both public opinion and ultimately Community policy. This not only includes groups such as Greenpeace and Friends of the Earth: it includes groups meeting in an ad hoc manner such as the North Sea Ministerial Conferences, or more formally the Ramsar Convention for the protection of wetlands. Most North Sea countries are party to the Convention.

For example, three international conferences on the protection of the North Sea (ICNS) have taken place (1984, 1987, and 1990). The last of these dealt with coastal zone management as its main topic.

What is perfectly clear is that throughout Europe there is a confusion of coastal users, administrators, interest groups, legislators and managers which renders the achievement of an acceptable all-embracing management policy very difficult indeed.

The Community has issued a Working Document concerning the special needs of the coastal zone (10 October 1990) partly because it recognised that earlier positions have not been as effective as some would consider that they should have been. These include various Community Environment Action Programmes and the European Coastal Charter (which was supported by a Resolution of the European Parliament on 18 June 1982).

The Commission is currently preparing proposals on strategic environmental assessments which will ultimately provide the broad framework within which more detailed requirements of coastal zone management will operate.

In addition, the Commission proposes to establish a legal instrument on the planning and protection of the Community's coastal zones in order to

(a) reinforce coastal land use and sea use planning procedures in Member States
(b) avoid threats to the environment as a result of economic development.

This instrument, if enacted, will require Member States

(a) to put in place a Coastal Zone Management Plan
(b) to legislate in order to ensure that all land use development within the coastal zone accords with this Management Plan
(c) to ensure that this Plan balances the demands of economic development and the conservation of natural resources
(d) to identify environmentally sensitive coastlines (by means of an inventory)
(e) to establish Coastal Protection Schemes for all environmentally sensitive coastal zones and areas interacting with them
(f) to submit physical planning projects within the coastal zone to an environmental impact assessment.

Conclusions

There have been many different reactions to the management of coastal problems world-wide. Not all of these form an appropriate parallel to the UK situation. For example, to a considerable extent the legal backing to coastal administration in The Netherlands is a function of this physical environment and the need to protect large parts of the country from inundation by the sea. A protective function therefore dominates both thinking and the legal context of coastal planning.

This raises the interesting point that parallels in administrative response to the needs of coastal management should really be sought between countries that are facing similar physical coastal circumstances. The Dutch example may be appropriate for consideration in the UK context, but it may be more applicable to selected parts of the UK coastline (i.e. those which face similar problems to the one faced by The Netherlands) rather than to the coastline as a whole.

In terms of trends in coastal management legislation the most obvious trend is the change in emphasis, since the 1970s, to protect the coastal environment in general and unique ecological habitats in particular. However, legislation is frequently directed at elements of the coastal zone (e.g. sand and gravel removal) rather than the coastal zone management system as a whole.

In a multi-faceted environment such as the coastline, with its multitude of pressures for different uses, it cannot be a balanced and long-term solution to devise legislation which addresses one aspect (or only a few aspects) of the whole range of pressures and ignores the rest. In time there would be a need to provide a counter-balance in the form of new legislation.

In the case of The Netherlands there is a slightly different emphasis in that the certain elements of the coast (e.g. the dune system) offer a vital protection to much of Holland from flooding. Preserving the dune system is therefore a matter not only of ecological interest: it is also one that is necessary for self-preservation.

In the case of Spain there is a clear fear (expressed through the Coastal Law of 1988) that, especially on the Mediterranean coast, uncontrolled development has gone too far. The rapid drive to meet a booming tourist demand, with its potential for generating a large amount of income, is now seen to have involved the destruction of a substantial part of the national coastal heritage. Steps have been taken to swing the pendulum back towards a more environmentally-sensitive approach to the management of the coastline.

In very few countries is there any effective legislation to deal with both the land and its near-shore waters in one planning instrument. New Zealand is one of the very few countries in which provincial land use planning schemes can be extended into the sea, though there are only one or two examples of it actually being done. In Australia some of the states have also included coastal waters within their planning legislation.

Since, in a physical (dynamic) sense the coastal system includes a delicate inter-play between land, shore and sea, it would seem appropriate for each to be considered within one competent and relevant management unit.

Legislation in some places (e.g. USA, parts of Australia, France) is now attempting a fuller integration of the needs for economic development along the coast and the pressures for coastal environmental preservation/conservation. However, the thought of a national administration having planning power over a local administration, especially after a period of administrative devolution away from the national towards local government, is politically no longer easy to accept.

Administration of the coastal zone is made very difficult when several Acts exist within the legislation that all have a direct bearing on the coastal zone. This is the case in most countries, with different Acts to deal with fisheries, environmental quality, pollution and the dumping of waste, merchant shipping, and local planning and development.

Following the principles laid down in the Brundtland Report *Our Common Future* (1987) there is a need to protect the "special" areas (including the environmentally sensitive and ecologically valuable ones) and refrain from causing environmental damage in the rest. Although there is a need to accept that coastal economic activities will continue, including living near the coast, it is nevertheless desirable to avoid significant coastal degradation.

Of all of the overseas countries it is perhaps the USA which comes closest to providing a role model for the UK. In this context it is interesting to note that the declared US position recognises that

> "There is a national interest in the effective management, beneficial use, protection, and development of the coastal zone" (Coastal Zone Management Reauthorization Act of 1985).

Such a statement leads to the view that policy concerning the administration of the coastal zone will have to be set at national level, with more local participation as appropriate.

This emphasis on the national importance of the coast is reinforced in the 1988 edition of the *United States Code*, produced by the Government Printing Office in Washington. Chapter 33 deals with Coastal Management, and sets out the declared findings of the US Congress after debating the problems associated with the management of the coastline. These findings amount to a statement of policy regarding coastal management that recognises the special value of the coastline to the nation as a whole, and the opportunities for development that the coast provides.

"The coastal zone is rich in natural, commercial, recreational, ecological, industrial, and esthetic resources of immediate and potential value to the present and future well-being of the nation."

The USA examples are relevant, transferable, and useful in that they show that

(*a*) national responsibility has been assumed for general policies concerning the management of the coast (federal responsibility)
(*b*) local plans can be set within an overall management strategy (state responsibility)
(*c*) coasts are seen to be dynamic systems with a value to the nation as a whole because of their intrinsic ecological, aesthetic, cultural and other properties
(*d*) management through planning should include a consideration of the whole range of coastal assets and alternative uses (the corollary to this is that it makes the adoption of a major engineering strategy less likely)
(*e*) national funds are used in support of necessary and approved management strategies
(*f*) the needs of basic research and education are provided for
(*g*) consultation is encouraged between policy makers and administrators at all levels from the national to the local, and public participation in discussion is also encouraged.
(*h*) local flexibility and responsibility exists within a national strategy.

If lessons can be learned from this examination of overseas practice they have to include the recognition that

(*a*) the acceptance of the fact that no country has found an "easy" solution to the complexities of coastal management
(*b*) there is a general recognition that national government has to carry the responsibility for the creation of a national coastal management policy
(*c*) the boundaries to the administrative units which happen to include a portion of the coast may not be coincident with the management unit required for effective coastal management
(*d*) too many problems occur when local governments take unco-ordinated piecemeal action along separately administered sections of the coastline, and their actions need to be contained within an overall general coastal management strategy.

Acknowledgements

This paper is based, in part, on a report presented to the Department of the Environment as part of the contract let to Rendel Geotechnics on *Planning Policy for the Coast, and Earth Science Information in Support of Coastal Planning*

PLANNING AND MANAGEMENT

and Management. Helpful discussion with E. M. Lee (of Rendel Geotechnics) and with members of the DoE Steering Committee (for this contract) is gratefully acknowledged.

Discussion

The pressures on the coastal zone were well illustrated by a description of problems in Hampshire and Dorset relating, inter alia, to population, tourism, leisure activities (particularly sailing), shipping, pollution and conservation. As a first step Hampshire County Council had produced a report giving their views on a strategic approach. Dorset were producing a strategy for their coastline while the SCOPAC coastal group was investigating the application of Coastal Zone Management (CZM). Consultation was proposed for developing a long-term coastal defence strategy in the area and for a strategy for marinas and moorings in the Solent.

It was accepted that the PPG was limited by the existing planning system and that changes in planning laws took a long time. Nevertheless the general view was that the PPG fell far short of what was necessary to achieve effective CZM and that a totally new approach was needed. The Report of the Environment Committee of the House of Commons on Coastal Zone Protection and Planning was welcomed particularly as the conclusions might influence the final edition of the PPG, once the report had been considered by the Government.

Comments on the needs of coastal planning included the following points

(a) planning should not stop at low water mark
(b) coastal processes demand a strategic approach
(c) CZM requires a national strategy
(d) to achieve co-ordination a lead authoritty must be defined
(e) there must be control over activities as well as land use
(f) although the advantages of consultation were understood it could not be relied on without legal backup
(g) detailed planning, conservation and engineering issues must be considered
(h) CZM is about power to execute and about money, including compensation for roll-back

Can the DoE and MAFF get together for a holistic approach?

The difference between planning and management was emphasised. An example was given of a strategic plan which had failed because no-one saw it as their responsibility. Successful plans had usually had a lead authority to manage.

DISCUSSION

In defence of the planning system it was pointed out that the UK was a pluralistic society and planning integration must be seen to be at a local level. There was now a search for a mechanism to resolve conflicts in the coastal zone but national aims might not be a solution. If interested parties were reluctant to get together informally, legislation might not solve the problem. The issue of LWM was difficult to resolve as it was tied up with administrative and political boundaries. CZM was a complex problem which required a complex answer. It would not be as simple as some delegates suggested.

It was pointed out that coastal groups now covered about 98% of the coastline of England and Wales. Some had arisen from local initiatives and some from initiatives by MAFF. It was suggested that Australia might be a better role model for CZM than the USA. The New South Wales environment plan was implemented in 1975-76. Local authorities remained responsible for planning but had to consult the PWD on any work in the coastal zone.

16. Coastal zone planning

T. M. COX, Borough Planning Officer, Sefton Metropolitan Borough council

The coastal zone

The coast is the interface between sea and land. Without human intervention it is an area of pressure, a dynamic and constantly changing environment, which, in geological terms, is ephemeral. Human beings, however, bring a shorter time scale and add to pressures exerted by natural processes as well as having sometimes locally profound effects. The coast is an area of competing activities: houses, recreation, industry, ports, waste disposal, agriculture, fishing, mineral extraction and natural habitats are perhaps the most obvious. The interactions are complex and the conflicts at times only too obvious.

Human activity results in a need for coast defence and protection works. Changes in the balance of natural forces can have dramatic and costly results. The loss of "coastal quality" - scenic and wild life habitat, both above and below the water mark - has become a major issue over the last 30 years. The scale of destruction is only too obvious in Europe, especially round the Mediterranean.

The purpose of "coastal zone planning" is, therefore, to find a means whereby spatial conflicts and environmental impacts can be assessed and balanced, and judgements taken, thus creating a framework for effective management. Carter (ref. 1) suggests that there are three approaches to coastal management. The first regards it as "an undemanding 'Friday afternoon task'. 'Off the shelf' solutions are applied, for problems ranging from coast erosion to habitat conservation. The second sees an increasing appreciation of coastal issues, although the sheer magnitude of many problems, embracing a wide variety of impacts often operating on different time scales, may be offputting. The third involves a deeper understanding, in which coasts are viewed as dynamic systems linking physical, chemical, biological and socio-economic processes." Reaching such an understanding and determining consequent planning and management frameworks is a daunting task.

But what is the coastal zone? Archer and Knetch (ref. 2) define it very broadly as including areas both landward and seaward of mean low water. Jones (ref. 3), for example, defines it as "a limited area varying in width according to physiographic conditions with a seaward boundary defined by low mean water mark". This last is primarily dictated by the jurisdiction of local authorities. Others, such as Gubbay (ref. 4), would argue, with some justification, that this is too restrictive and fails to recognise the importance of the marine environment and the interaction of land and sea which is the essence of the coastal zone.

However the coastal zone is defined, the need for effective planning and management is becoming more urgent. Participants at the recent European Workshop on Coastal Zone Management (ref. 5) considered that

> "the European Coast is now subject to unprecedented change through a combination of natural process and man induced impacts"
>
> "present political, administrative and management systems have often failed to prevent damaging development and have not managed the environmental consequences".

A dramatic message perhaps, but a concern echoed by many organisations who, for example, have given evidence to the Commons Environment Committee in the UK context.

The British experience

Coastal areas of the UK face a variety of pressures which give some urgency to a more effective regional and national approach. While the list of issues outlined in "Planning for the Coastline" (ref. 6) remains as relevant 20 years on as it was in 1970, it is depressing that only one of its recommendations has survived and prospered - that of heritage coasts. A current list of concerns might include

> (a) rising sea level and changes in weather patterns with consequent impacts on erosion and increased risks of flooding
>
> (b) major estuarial and coastal developments including tidal power installations
>
> (c) the effects of offshore mineral extraction, oil and marine aggregates
>
> (d) marine pollution and sea water quality
>
> (e) continuing losses of significant wild life habitats.

Some of these problems have been well documented in the Nature Conservancy Councils's "Nature conservation and estuaries in Great Britain" (ref. 7), the RSPB's campaign documents "Turning the tide" (ref. 8) and "A future for the coast" (ref. 9), published by the Marine Conservation Society for the World Wide Fund for Nature.

Coastal zone planning is not a new concept. The Dutch, not surprisingly, have had a sophisticated system for many years but other examples of a comprehensive approach are difficult to find. The US Coastal Zone Management Act of 1972 is well known and all 35 coastal states have now participated. Most now have detailed coastal land use zoning, habitat inventories, access guides and resource evaluations (ref. 10). End results have been variable, but much innovation has taken place. The applicability of American experience is, however, limited by different approaches to funding, the patterns and rights of land ownership and jurisdictional conflicts at various government levels.

In the UK there is, of course, no comparable 'top down' approach and policy has been incremental, characterised, with the notable exception of heritage coasts, by a series of local initiatives and the influence of voluntary bodies, particularly the National Trust, not simply in lobbying but in active management. The land use planning system has had some success, but its potential effectiveness has been hampered by a lack of perception of coastal issues, both locally and nationally.

A review of coastal planning authorities undertaken for the RSA's coastal conference in 1991 (ref. 11), highlighted the fragmentary nature of activity, with most counties pursuing limited specific initiatives. The honourable exceptions are well known, particularly Hampshire. This view was reinforced in a paper to a recent regional conference (ref. 12). A survey of coastal counties revealed that only 2 had an approach to the coast as a whole and 5 had no coastal initiatives of any sort. At District level, 18 had planned initiatives and 20 had no involvement.

Another initiative is that of SCOPAC (Standing Conference on Problems Associated with the Coast) - a group of nineteen local authorities, with harbour authorities, the NRA and the other agencies which aims to ensure a co-ordinated approach to all coastal engineering works and related matters between neighbouring authorities on the coastal sections on the south coast of England. The range of issues addressed by SCOPAC goes well beyond strict engineering concerns and perhaps illustrates the potential role of the fourteen coastal engineering groupswhich have been established, mainly by District Councils, and which now cover most of the English and Welsh coastline. The Anglian Region of the NRA has pioneered a comprehensive approach to coast erosion problems between the Thames and the Humber, drawing in local authorities and other agencies.

While these are developments with much promise, the British approach remains very British. It is fragmentary, inconsistent (from market leaders to no activity), assumes problems will be recognised by sensible people, depends on a substantial degree of voluntary co-operation, relies heavily on voluntary or special interest groups, is characterised by a large number of overlapping jurisdictions and lacks any coherent framework at local, regional or national level. But despite these criticisms, it does have its achievements. Heritage coasts are the outstanding example, and it can be argued that, in terms of government policy, there is a trend from disinterest to mild concern.

PLANNING IN RESPONSE TO CHANGE

The importance of local initiatives is thus a key part of the British experience - a 'bottom up' approach arising from examples of active coastal management and small scale action. The Sefton Coast illustrates this progression with a coast management scheme leading to Coastal Zone Planning.

The Sefton Coast

The Sefton Coast forms an identifiable geographic unit between the Ribble and Mersey estuaries on the North West coast of England. The origins of its settlements relate to the sea from the growth of Liverpool as a port in the south, to that of Southport as a resort town in the north. Settlements developed on wide areas of blown sand along the shore, avoiding the peat mosses which lay beyond. Much of the agricultural land to the east lies below mean high tide level. The coast has no inherent strength and is affected by the tidal regimes of both estuaries, which determine the transport of sediment up and down the coast. With a tidal range of 10.3 m and storm surges which can add 0.75 m to high tide levels, the coast is in a delicate state of balance and is easily influenced by human activity.

While there had been sporadic concern with coastal matters throughout this century, for example when houses at Crosby fell into the sea in the 1920s, it was not until the 1950s that academic interest became more prominent (ref. 13), and not until the 1970s that this translated itself into a concern other than immediate sea defences. It was in the 1960s that the National Trust and the Nature Conservancy brought substantial areas of land on the North Irish Sea 'soft' coast at Formby to protect its scenic value and/or wild life habitats. The damage being caused by unrestricted access was only too obvious in

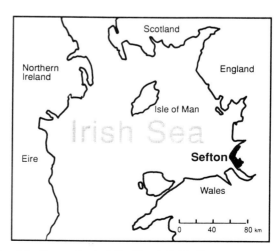

Fig. 1. Sefton Coast in the context of the North Irish Sea

some areas, with expanding patterns of erosion caused, in part, by trampling of marram grass, so weakening the dune front and destroying valuable habitats.

While Lancashire County Council had looked at some of the area in the 1960s as a possible Country Park, the Local Authorities had taken little real interest until 1977 when the submission of a planning application for a golf course in the dune land south of Formby triggered a whole series of events leading to the formation of the Sefton Coast Management Scheme (ref. 14). This scheme has been well documented; the point here is to demonstrate the link with coastal planning policy.

The Sefton scheme originally covered some 14 miles of dune coastline between the major settlements in the north and south of the Borough. A project officer was appointed who provided the drive and co-ordination, and the operation was assisted by a relatively simple pattern of land ownership (Sefton Borough Council, the National Trust, the NCC and the Territorial Army).

The early years were characterised by major land restoration projects needed to deal with rapidly eroding areas which had resulted partly from recreation pressure and partly from the storminess of the 1960s. A Management Plan was developed setting out a framework for action by the individual managing agents and a Ranger Service was established by the Borough Council working closely with the National Trust and NCC wardens. The role of volunteers, both in carrying out work and supporting the ranger service, was, and continues to be, a key feature. The Management Plan was revised in 1985 (ref. 15) with more emphasis on long term monitoring, interpretation and education.

Thus, the management of the coast has gradually developed in greater depth and demonstrated that large numbers of people could enjoy a delicate environment in a way that protected and enriched key parts of it. Meanwhile, concern with the wider context was growing. The management scheme had always had strong links with the planning process, but the planning policy framework was rudimentary. Moreover, the management scheme only covered part of the coast. Could the same principles be applied to the remainder, much of which was built up? Would a stronger framework ensure that the management scheme itself was more robust? These questions arose out of very detailed planning exercises being carried out on the future of Southport's leisure zone as well as issues surrounding the port of Liverpool.

As well as local concerns, outside influences had an important part to play. From the mid 1980s onwards, the coast increasingly became a matter of national concern, particularly as development and tourist pressures mounted. The possible impact of global warming began to acquire a popular following. Energy developments multiplied further. Sea use management and sea water quality became more prominent issues, with increasing interest from Brussels. Major studies and conferences on the Irish Sea and North Sea took place. The environment in its widest sense achieved, during the decade, a far more prominent, and hopefully enduring, position among local, national and international issues.

In a small way, all these concerns focused in, for Sefton, on the 23 miles of coastline. The only statutory policy vehicle available to the Local Authority formed part of the land use planning process - an important but hardly all encompassing mechanism. It cannot, for example, deal with issues and problems below low water mark. It was, however, important to underpin the work of the previous decade by something which has at least some 'legal teeth', the Borough's Unitary Development Plan. It was also felt particularly important that a Coastal Zone should be identified geographically and land use and development planning policies established for that area.

Coastal planning policies

Planning policy approaches which identify and respond to the coast as a major issue are not easy to find. Hampshire (the market leader), in its County Structure Plan (1991), sets out five policies for the coast and the County's approach is developed in "A Strategy for the Hampshire Coast" (ref. 16). A review of local development plans on either side of the Irish Sea (ref. 17) produced scant references to coastal planning, although there is much evidence of coastal pressures, specific initiatives, and some concern, particularly in Wales and Ireland, at the loss of coastal quality. Norfolk has recently introduced policies into its draft Structure Plan, not without difficulty, to prevent development in potentially vulnerable areas as a response to rising sea level and climatic change. Northumberland has produced a draft Coastal Management Plan for the whole of its coastline (ref. 18). It is, however, non-statutory and "does not necessarily reflect the views or policies of all partner organisations". There was thus little available particularly at District level, to act as a model for Coastal Zone policies in Sefton's Unitary Development Plan (ref. 19).

A relatively narrow coastal zone was defined, varying on the landward side, from as little as 100 m to 2.5 km, stretching down to low water mark. The coastal planning zone includes all the undeveloped dune and marsh areas, excludes existing residential areas but includes the Docks and the Southport leisure area. Three policy objectives were set out under the following heads

(a) physiographic conditions - to require that the natural conditions and processes on the coast are recognised and taken fully into account in any development proposals, and to restrict development, where applicable, for physiographic reasons

(b) natural resources - to conserve, and where appropriate, restore and enhance the natural resources within the coastal zone, including mineral and marine resources, natural and semi-natural coastal fauna and flora and their habitats, and the amenity values inherent in the coastal landscapes

(c) coast dependent uses - to promote coast dependent economic development and renewal, including the restoration and enhancement of unique coastal areas and structures compatible with the surrounding open areas.

Objective (c) is the corollary of objective (b). If the undeveloped parts of the coast are to be given their proper value and protected, then existing coastal development should be equally valued. A port or seaside resort is as much a resource in its own way as a saltmarsh or dune belt. The approach adopted thus tries to encapsulate an holistic approach to coastal planning.

Three of the 17 policies arising from these objectives are of particular importance. The first attempts to ensure that the coast is safeguarded as a natural sea defence, and to avoid unnecessary or unjustified costs arising from the need to protect new development against the sea. The second sets out the sort of information the Council would require in an environmental statement to accompany any development falling within the scope of the Assessment of Environmental Effects Regulations 1988, covering sediment transport, erosion and accretion, coast protection/sea defence, the effects on the nature conservation value of the coast and its landscape and scenic qualities. The third underpins the Coast Management Plan by seeking to prevent any development which would conflict with its aims. In public consultation on the Plan, a wide measure of public support was received for both the general intent of these policies and their more detailed content.

Policies have been designed to ensure that any development in the zone is linked with the natural systems which bear upon the area as a whole. Seawards, the concern is with changes in sediment deposition which affects beach quality and coast protection. Landwards the links are concerned with coastal quality and access. Most importantly it attempts to create a strong bond with the management regimes established in the Coast Management Plan (CMP) and the day to day actions for which the CMP provides the framework.

The Sefton Coast, in common with others, is part of a highly dynamic system subject to seasonal cycles of erosion and deposition as well as longer term changes in weather patterns, climate and sea levels. While there continues to be some controversy about the timescale associated with rises in sea level, even a moderate rise, linked with increased storminess for example, could lead to severe damage. The precautionary principle is therefore of considerable importance. Underdeveloped coastal areas provide a valuable

PLANNING IN RESPONSE TO CHANGE

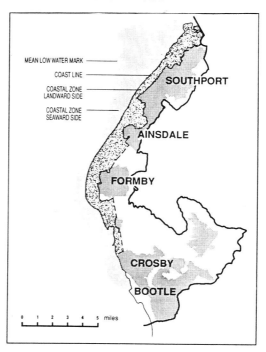

Fig. 2. Sefton - Coastal Planning Zone

buffer between the sea and the hinterland. Definition of a coastal zone and policies supporting it are thus an important means of ensuring that the 'buffer' role continues. Here it is important to stress the role of engineering colleagues in their coast protection work which provides the data base for supporting such policies and, hopefully, winning planning appeals.

Planning Policy is, of course, only part of a spectrum of coastal planning and management measures ranging from detailed site management actions to international conventions. What is important is that problems are identified and tackled in a context which recognises inter-relationships and impacts, and which does not see a problem as simply one of sea defence, or habitat protection or accommodating visitors. The 'grassroots' evolution of policy, as Jones (ref. 20) points out, is particularly beneficial because it sharpens policy relevance and strengthens its credibility. There is plenty of this type of experience in the UK on which to base a broader approach.

The Sefton experience is very much an example of the 'bottom up' approach. Despite its long development and strong local support there is also a consciousness of the fragility of what is, for the most part, a discretionary activity, should political support wane, key individuals leave or resources be cut. It has also become clear that, even with a relatively extensive coastline, action, or in some cases lack of action, beyond Sefton's boundaries - for example, the tidal barrage proposed on the Mersey - can have profound effects. These wider impacts are increasingly recognised. Studies such as that undertaken for the Anglian Region of the National Rivers Authority, and the operational clustering of coast defence authorities in 'coastal cells' are, for example, welcome developments. It is essential, though, that these engineering-led initiatives are effectively followed up, supported by or operated in the context of effective planning policies.

Integration

Management experience in coastal areas has one clear message: the need for integration - a real pooling of effort which reflects at professional and political level the rich interlocking of concerns and actions. Professional interests are helpful in the skills they provide but unhelpful in the barriers that can and do emerge. Active interdisciplinary work is needed which does not seek to define an exclusive role for the ecologist, planner or engineer but allows each to appreciate the contribution of others in constructive and productive interaction.

Professional integration, however determined, is not enough. Numerous commentators have noted the difficulties which arise from the multiplicity of agencies. Parish (ref. 21) for example points out that the seaward boundaries of local authorities lie along the low water mark - fairly straightforward in seafacing coasts but much less so in estuaries and tidal rivers. Seaward the position changes to a system of functional and vertical controls exercised by central Government. Gubbay (ref. 22) estimates that 25 different departments and agencies are involved beyond low water mark. There are thus considerable differences in the way the coast is planned and managed either side of the low tide mark. It is, at the simplest level, often difficult to know what is actually going on. The local authority associations in their evidence to the House of Commons Environment Committee (ref. 23) argue for extending statutory planning controls to inshore waters.

Statutory planning controls are, however, not without their problems. For example, permitted development rights to Dock and Harbour Authorities allow a wide variety of activities which can be damaging to the environment both landward and seaward. The problems of dust blow from coal storage in the Liverpool Docks (ref. 24) and the impacts of large Wightlink Ferries between Lymington and Yarmouth (ref. 25) are two examples which are dissimilar.

What is badly needed is both a national and local framework which recognises the urgency of the situation and provides the direction and impetus for proper planning and management. It was a considerable disappointment that the White Paper "A Common Inheritance", while mentioning issues such as sewage sludge disposal and the extension of marine conservation areas, did not reflect the need to provide a strategic framework for the coast as a whole.

One limited, but hopeful, sign is the likely emergence of a Planning Policy Guidance Note on Coastal Planning from the Department of the Environment, much anticipated for some months. While this is unlikely to usher in a new age of Coastal Zone Planning, recognition of the coast and coastal issues by Government in a 'planning context' is particularly welcome. What it will not include has been well publicised. It does not seem likely to herald any major reorganisation of coastal responsibilities, extensions to the jurisdictions of local authorities or anticipate requirements for formal groupings or statutory coastal plans (ref. 26). On the positive side, however, it should encourage local planning authorities to look at coasts as a whole in relation to a complex of issues and may well provide, with the new flood risk circular, guidance on the approach to risk definition and assessment.

The way ahead

Much that has happened in the last four years has pointed to the broad framework that should be adopted in this country. Many local initiatives have provided a valuable store of detailed site specific experience on which to draw. The whole picture, however, is of sporadic activity - a reliance on local voluntary activity or discretionary action by local authorities. The major weakness is primarily at central Government level in failing to provide a framework which ensures that the sum total of local activity and experience is translated into a consistent and relevant approach for the UK's coastal areas. The Countryside Commission, in revising its policy on Heritage Coasts (ref. 27) makes this point strongly and it is one which has been repeated by many of those making representations to the Commons Environment Committee.

While the Government's preparation of a Planning Policy Guidance Note is welcome, as Gubbay (ref. 28) points out, this is hardly sufficient to maintain momentum for coastal zone management in the UK and will not adequately link policies between the landward and seaward side of the coastal zone.

The conclusions of the local authority associations (ref. 29), submitted to the Environment Committee, are well worth repeating here. The associations argue that

(a) coast protection and planning suffers from inadequacies in current legislation, divided responsibilities and a general lack of co-ordination
(b) there is a plethora of bodies involved in coast protection and planning including Crown Estates, the NRA, Water Companies and over 30 Government departments and agencies; while there are examples of good

co-operation between agencies in some areas, the lack of any statutory back-up means that in many areas there is nothing
(c) there is little strategic thinking on coasts or estuaries
(d) there is no comprehensive legislative framework for the marine environment to parallel that covering the land in the form of the Town and Country Planning Acts.

An energetic approach to these problems is essential and the consistency of view between a wide variety of organisations is striking. That approach must come from central Government. The voluntary sector and some local authorities have shown the way. Again, the local authority associations argue, among other things, for

(a) a review and rationalisation of responsibilities in the coastal zone with the attention of statutory planning powers below low water mark
(b) Government encouragement, guidance and support for strategic planning for the coast including estuaries within a national perspective
(c) a standing forum for each stretch of coasts with a designated lead authority.

The example of the Sefton Coast does indicate that it is possible to develop coastal zone planning as a valuable part of coastal zone management but that it is a slow process, particularly fragile and inevitably partial. The establishment of coastal 'cells' by engineers and wider initiatives such as the Mersey Estuary Management Plan all speak of a growing recognition of wider groupings and inter-connection of many activities and disciplines. However, it is too slow, sporadic in its coverage, is not robust, provides no national framework and is potentially wasteful of time and skills.

Action is needed in the UK, and indeed elsewhere in Europe, to ensure that effective planning of the coastal zone does take place on a consistent basis and within similar time scales. Enough has been done to demonstrate that there are practical approaches and solutions which linkboth land and sea; policies which do not deal with sectoral interests in isolation and are supported at national, regional and local level. This will need

(a) national commitment and a requirement to produce coastal zone management plans reinforced through statutory planning systems
(b) a recognition of the key importance of a national framework implemented through locally based groupings with clearly assigned lead responsibilities
(c) funding regimes which require, as a pre-requisite for financial support, a planning and management framework of the type suggested
(d) a review and simplification of agencies and responsibilities affecting the coastal zone

(e) the effective involvement of non-governmental bodies who have much expert knowledge and passion to contribute.

The preamble to the European Coastal Charter (ref. 30), developed by a grouping of European coastal nations in the early 1980s, sums up the reason why coastal zone planning is so important. Action, it suggests, is increasingly necessary to reconcile both necessary development with the protection of the ecological and aesthetic value of the coastal zone. The planning needed must cover organisational as well as implementation issues. It is very likely that a similar principle will be incorporated in a forthcoming European Community directive on coastal zone management.

Central Government must give a clear lead at national level on coastal zone planning and management. It could thus be in the forefront of European action. As a maritime nation this is surely a role we should occupy.

References

1. CARTER R.W.G. Coastal zone management. Comparisons and conflicts. Planning and management of the coastal heritage, edited by Houston and Jones, Sefton MBC, 1990, 45-49.
2. ARCHER J.H. and KNECHT R.W. National Coast Zone Management Programme - problems and opportunities in the next phase. Coastal Management, 1987, vol. 5, 103-120.
3. JONES C.R. Planning in the coastal zone.
4. GUBBAY S. Management for a crowded coast. Landscape Design, Dec. 1991/Jan. 1992, 10-12.
5. EUROPEAN WORKSHOP ON COASTAL ZONE MANAGEMENT. Communication from participants at the workshop held at Poole, Dorset, 1991. See also COUNTRYSIDE COMMISSION. Europe's coastal crisis - a co-operative response. Countryside Commission, Cheltenham, 1992.
6. COUNTRYSIDE COMMISSION. The planning of the coastline. HMSO, London, 1970.
7. DAVIDSON N.C. et al. Nature Conservation and Estuaries in Great Britain. Nature Conservancy Council, Peterborough, 1991.
8. ROTHWELL P.I. and HOUSDEN S.D. Turning the tide. RSPB, Sandy, 1990.
9. GUBBAY S. A future for the coast? Proposals for a UK Coastal Zone Management Plan. Report to the WWF, Marine Conservation Society, Ross on Wye, 1990.
10. CARTER R.W.G. op cit, 47.
11. KING G. Planning: a local authority viewpoint. Unpublished paper to RSA Conference "The Future of Britain's estuaries", London, February 1991.

12. BROOKE J. Coastal zone management. Unpublished paper to NW Royal Town Planning Institute Conference on coastal planning and management, Chorley, November 1991.
13. GRESSWELL R.K. Sandy shores of South Lancashire. University Press, Liverpool, 1953.
14. COX T.M. Coastal planning and management in a metropolitan area. Planning and management of the coastal heritage, edited by Houston and Jones, Sefton MBC, 1990, 32-36.
15. COX T.M. Coastal management plan and work programme. Sefton MBC, Bootle, 1989.
16. HAMPSHIRE COUNTY COUNCIL. A strategy for Hampshire's coast. County Planning Department, Winchester, 1991.
17. SMITH H.D. and GEFFEN A.J. The Irish Sea - Part 4: Planning, development and management. University Press, Liverpool, 1990, 87-97.
18. NORTHUMBERLAND COUNTY COUNCIL. Northumberland Coast Management Plan - Consultation Draft. Environment and Economic Development Department, Morpeth, 1991.
19. COX T.M. A plan for Sefton - Unitary Development Plan (Deposit Draft). Sefton MBC, Bootle, 1991.
20. JONES C.R. Planning in the coastal zone, op cit, 6.
21. PARISH F. Coastal management. The practitioner's viewpoint. Planning and management of the coastal heritage, edited by Houston and Jones, Sefton MBC, 1990, 29.
22. GUBBAY S. Management for a crowded coast. Landscape Design, Dec. 1991/Jan. 1992, 10.
23. LOCAL AUTHORITY ASSOCIATIONS. Coastal zone protection planning. Memorandum of Evidence to the House of Commons Environment Committee, London, October 1991.
24. SEFTON MBC. Memorandum of Evidence to the House of Commons Environment Committee, Bootle, October 1991.
25. WILDLIFE LINK. A selection of case studies illustrating the need for coastal management in the United Kingdom. Unpublished 1991.
26. BACH M. Contribution to NW Royal Town Planning Institute Conference on coastal planning and management, Chorley, November 1991.
27. COUNTRYSIDE COMMISSION. Heritage coasts: policies and priorities. Countryside Commission, Cheltenham, 1991.
28. GUBBAY S. Coastal zone management. ECOS, Vol. 12, No. 2, 1991, 6.
29. LOCAL AUTHORITY ASSOCIATIONS. Coastal zone protection and planning, op cit, 12-13.
30. CONFERENCE OF PERIPHERAL MARITIME REGIONS OF THE EEC. European Coastal Charter.

17. Shoreline management: a question of definition?

I. H. TOWNEND, Sir William Halcrow & Partners Ltd

Recent developments in shoreline management have led to moves towards a more holistic approach based on a responsive management framework. This in itself parallels developments in coastal management and raises the question as to whether the distinction between the two is useful and warranted, or not. A framework for shoreline management is outlined, which makes full provision for understanding and responding to natural processes and attendant change. The relationship of this framework to the wider remit of coastal management is then discussed in order to establish the similarities and differences and consider how the two might co-exist.

Introduction

Pressure for coastal space and coastal resources has steadily increased over the centuries. Initial settlements on the coast took advantage of the food and living areas available. Travel by sea then provided a link between coastal settlements. This mobility did, however, also lead to the need for defences to guard against possible invasion. As trade expanded large ports and cities developed, often acting as focal points for economic growth. The advent of the steam train gave rise to seaside holidays and the relatively new industry of tourism, bringing with it additional pressures on the coastal resource.

The area of coast available is, however, finite and there is therefore a danger that the various forms of coastal resources could be over-exploited. The risk of loss or conflict is exacerbated by the dynamic and sensitive nature of the interface between land and sea. This requires that we endeavour to work with rather than against nature, implying that our efforts to manage the coast must be based on proper understanding of the processes.

Coastal management

If it were not for our desire to make use of the coast and its resources, there would be no requirement for coastal management, so the following relatively simple definition of coastal management seems particularly appropriate.

> "Our use of the coast must be organised so as to allow the reconciliation of the demands of DEVELOPMENT and the requirements of PROTECTION" (based on the European Coastal Charter: CPMR, 1980)

PLANNING IN RESPONSE TO CHANGE

This definition recognises that conservation and rational development must co-exist if there is to be the economic benefit to sustain protection of the coast in the long term. The coastal resource base comprises bio-physical, socio-cultural and economic elements, all of which interact. Organisation of these elements implies planning, which must be sufficiently detailed to provide a national, regional and local framework, but with the flexibility to be adopted to the particular physical, economic and legal circumstances of the country or region in question.

Any attempt at coastal management must therefore consider a wide range of issues. As already noted the potential for conflict both between different coastal activities and with the natural system, demands a proper understanding of coastal processes. Indeed in some situations there are even conflicts within the natural system (e.g. habitat vs morphological form) which may need to be addressed. The framework for management must then embrace such things as legal, organisational and institutional issues; the preparation of management plans; the concept of zoning; the problems of regulation and enforcement; the need to develop public awareness and promote public consultation; and economic and social dimensions (refs 1 and 2). Such a multi-disciplinary coverage inevitably means that a wide range of organisations are involved, often each confining their interest to some small component of the whole (e.g. conservation, tourism).

Shoreline management

One such component relates to coastal hazards and the need to protect against either flooding, erosion, or both. Traditionally, the coastal engineer has provided a range of structural solutions to meet particular social needs. Over recent years there have been some fundamental changes to the provision of flood protection and measures to limit erosion. In practical terms this has resulted in new types of structure and a range of schemes which either emulate natural systems or operate in accord with the natural regime.

Implicit in such changes is a greater understanding of coastal processes and this has led to the study of whole coastal regimes, rather than just the frontage of interest. It soon became clear from the more regional studies, that not only can a more strategic approach to the provision of coastal works be adopted, but also a far wider range of options can be considered. This has given the incentive to develop suitable shoreline management techniques (refs 3 and 4). In this context we can define Shoreline Management as

Our efforts to direct and control the shoreline interface in respect of coastal hazards such as flooding and erosion.

As we shall see the wider scope that this definition implies, over the simple provision of coastal structures, means that shoreline management forms an integral component of coastal management and in particular must interface with the planning and regulation components.

This is illustrated in Fig. 1 (after Fischer, ref. 3), which shows the various stages in the shoreline management process, from the initial hazard to the feedback that results from experience. As well as efforts to control the shoreline there is a clear link to the wider considerations of shoreline use - so much so that any division may be artificial and perhaps undesirable. Any framework for shoreline management must therefore be consistent with both the direct objectives and the wider objectives of coastal management.

Management framework

The approach to shoreline management proposed is based first and foremost on a proper understanding of the processes and only then goes on to consider the management issues that result from potential uses of the shoreline. A fundamental requirement for both of these aspects is information. This must be gathered in such a way that it can be rapidly manipulated and analysed to determine the key processes which are in turn used as a basis for formulating the strategy.

There are three components to the management framework proposed, namely the Input, the System and the Strategy (Fig. 2). The principal Input is data, whether existing, generated by routine monitoring, or the result of

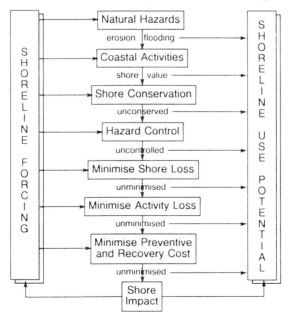

Fig. 1. Stages in the management process

forecasts. Extracting information from a wide range of archives rapidly generates a large volume of material. To provide easy reference, this needs to be adequately catalogued if the process of interpretation is to remain manageable. For a relatively short length of coast it may be that much of the information can simply be documented, possibly in report or tabular format. As the study area becomes large or more complex, so the use of a suitable computer database can be beneficial for the control and manipulation of data. In either case this should provide details of the coastal characteristics (morphology, ecology, usage, etc), the coastal forcing (waves, tides, etc) and the coastal response (erosion, accretion), which are used to develop the coastal classification. This can be conceived as a system for managing the information, i.e. a Management System.

As already noted the information on processes and usage provides a fundamental input to the development of a management strategy. Such a strategy must obviously be formalated within the constraints of the legal and institutional frameworks which prevail. It will therefore comprise objectives based on the legal obligations and aims of the organisation involved. These will be taken forward by policy guidelines which identify both the imposed constraints and strategic decisions which endeavour to make best use of available resources. Whilst the policy guidelines satisfy future planning, the management response options define how the policy can be implemented. Taken together the objectives, guidelines and response comprise the Management Strategy.

This framework can be initiated and developed to varying levels (ref. 4). It is also important to recognise that it is applicable to the management of any process based system. The Management Framework concept can therefore readily be extended from shoreline management into the realms of coastal management and river basin or catchment management.

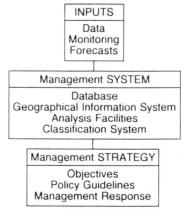

Fig. 2. Outline of management framework

Whilst a great deal of benefit is gained from moving to a regional strategy formulated on an improved understanding of the processes, this still does not acknowledge the dynamic nature of the problem. As well as man's actions on the coast, the environment itself is constantly adapting. There is therefore a clear need for some form of feedback loop within the Management Framework.

This is achieved by first collecting existing data to establish a baseline. A programme of routine monitoring then provides new Input. In a similar way studies of change can be initiated, such as a change of use, or changes due to forecast sea level rise. These can be carried out as independent exercises and the results compiled in terms of a new definition of coastal characteristics, forcing and response.

Both monitoring and forecast facilities are independent of the Managment System and require changes to components such as coastal response or coastal forcing to be specifically defined. Given a new definition of variables, the Management System can then be used to update the coastal classification. As our understanding evolves so the Management Strategy can be modified. Hence this framework provides a means of responding to changes, once they have been identified.

Management boundaries

When developing a formal approach to coastal or shoreline management the question of how to define the boundaries of the problem invariably arises. Unfortunately no single criterion is universally acceptable as an effective definition (Table 1). Whilst the coastal zone is the most widely recognised entity for management purposes, there is an alternative which derives from a systems approach to the problem. This is based on a coastal region, which exhibits a degree of independence from external elements and has the potential to be organised within an integrated framework. Vallega (ref. 5) develops the concept of such regions and outlines how they must be delimited to respect existing legal frameworks, minimise differences due to administrative jurisdictions and recognise the spatial extent of the coastal environment. This is of course wholly consistent with the much discussed holistic approach, seeking to optimise the use of resources working from the whole to the part. A fundamental requirement for a sustainable coastal region is therefore that it must recognise the true regional extent of the natural processes. This favours environmental boundaries (Table 1) and merits the additional effort needed to define them.

As Salm and Clark (ref. 6) noted, however, "to define a coastal management unit requires more pragmatism than theory, because the unit must be a realistic entity". Whatever means of delineating the area to be managed is adopted, the coastal margin must be regarded as mobile and therefore

PLANNING IN RESPONSE TO CHANGE

changing with time. Only when this is an integral part of the thinking behind planning and management, as well as an accepted public perception, can appropriate response strategies be adopted.

Input

The range of variables which must be covered is indicated in Fig. 3 under the three categories of characteristics, forcing and (the interdependent) response. This not only provides coverage of the relevant processes but also defines usage so that activities and possible conflicts can be considered when developing the management strategy (refs 7 and 8 provide further details of data coverage).

Given that the effectiveness of any management process is determined by the adequacy of the information upon which decisions are based, it is clearly important that the most essential variables should be up-dated on a routine basis. This means that whilst information on coastal characteristics will require periodic review, the more dynamic components will need to be collected as part of a structured monitoring programme. Such a programme

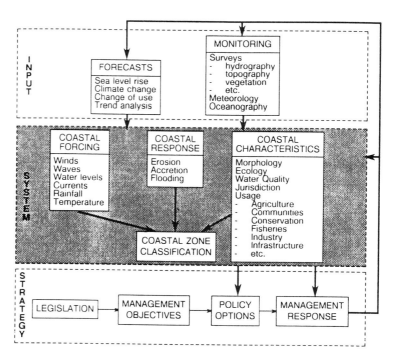

Fig. 3. Responsive management framework

Table 1. Criteria for defining management boundaries

CRITERIA	EXAMPLES	ADVANTAGES	DISADVANTAGES
Physical	Area defined by such things as: coastal plain, watershed and major roads on land; high water, low water and continental shelf edge to seaward	- simple to describe - easy to understand	- seaward extent difficult to establish - not necessarily consistent with natural processes
Administrative	national, country and district boundaries	- easy to understand - legally viable	- may be too restrictive - may artificially divide resources
Arbitrary	set back lines, Territorial waters	- ease of application	- arbitrary distance bears no relation to coastal topography, the natural system or economic activity
Environmental	selected units such as inlets, bays, reefs, mangrove etc	- sound morphological and/or ecological basis	- not easily understood - difficult to chart and legislate for

Table 2. Data requirements for forcing component

Description	Format	Frequency	Analysis	Use
Wind	Velocity and direction	Hourly	• Inshore wave climate • Offshore storm climate • Extremes	To assess shoreline exposure
Water levels	Height ref to datum	At least hourly	• Tides • Extremes • Joint wave and water level analysis	Required for inshore wave computations and allows significance of storm surge to be identified
Tidal Prism	Water levels and currents	Every 5-10 years	Estuary power curve	To define estuary dynamics and in the longer term to provide indication of changes in the hydraulic regime.

Table 3. Data requirements for response component

Description	Format	Frequency	Analysis	Use
Aerial Survey	Photographs	Annual	Foreshore and backshore levels	To identify key features, quantify change in both a plan and cross-shore sense and to monitor both physical and ecological changes
Hydrogaphic Survey	Depths along profile lines	Every 5-10 years	Changes with time	Establish limit of sediment exchange on shore face and provide detailed, up-to-date bathymetry for process studies
Land Survey	Levels along profile lines	Biannual	Changes with time	Allows changes in the beach to be studied and related to forcing data (Table 2)
Inspections	Text records	Biannual	Changes with time	Observation of beach features and coastal structures can be related to processes and key events, to help establish patterns of shoreline behaviour

PLANNING IN RESPONSE TO CHANGE

will focus on forcing components and aspects of the coastal response. Tables 2 and 3 (ref. 9) provide a typical scope for such a programme based on the need to collect relevant data without placing an undue burden on financial resources.

The implementation of well co-ordinated monitoring programmes on a regional basis places the activity of data collection on a formal and rational footing. This approach not only provides the input for the Responsive Management Framework outlined, but greatly improves the quality of design information and may also provide an early indication of the direct impacts of sea level rise.

Management system

At its most basic an information system provides access to data through some form of referencing. Traditionally, this is through an index system to an archive of records and documents. The alternative now available is to use some form of computer database to store and retrieve information. For coastal studies a convenient and relatively quick approach is to define a reference coastline and chainages along this line from some arbitrary start-point. All attribute entries into the database are then referenced to this fixed chainage system. This imposes a reduction on the data to either point or line type on the fixed reference line. Its advantage is that is readily allows spatial correlations in the alongshore dimensions and can be used to develop both maps and x-y plots showing values of variables against chainage along the coast.

The processes, the resources, and the uses we make of the coast all have a clear spatial dimension. By using geographic information systems (GIS), it is possible to describe real world objects in terms of their spatial description (point, line, area), their attributes (e.g. name, value, clsssification), and their relationship with other objects (i.e. topological relationship). As a basic data retrieval tool a GIS can be used to find where items are, what an item is, or to obtain a summary of all occurrences of an item. The real advantages are gained from the analytic capabilities. The tool can be used for classification, by comparing different data sets to seek patterns or combinations which suggest some form of relationship. By starting with known problems or phenomena one can adopt an inductive approach to seek general relationships. Finally it is possible to start with an accepted principle and see if this can be substantiated by the available data. In all such analysis, it is important to seek the underlying physical explanation for any relationships observed (usually by means of carefully focused modelling), and to be aware of spatial and temporal limitations of the data within the GIS.

Although GIS is still a relatively new tool in coastal management applications, it has been widely applied in a range of studies (ref. 10). These studies, however, only serve to illustrate the potential. For the proposed approach to succeed and justify both the initial set-up and subsequent maintenance costs, such tools must form an integral part of routine management. They must therefore be available for use by those responsible for the day to day management of the coast. Success will depend on well designed systems, appropriate training and a commitment to regional rather than local management.

Management strategy

The final component of the Management Framework is the Strategy itself. This must be formulated to function within the prevailing legal and institutional frameworks. Starting with clearly stated aims and objectives, the strategy establishes guidelines on policy issues and then sets out appropriate response options, acknowledging both the natural processes and the constraints imposed by the guidelines.

Aims and objectives

The prime aim of shoreline management (as defined above) is to provide hazard protection. In order to achieve this somewhat open ended goal it is necessary to define what is to be achieved in terms of more measurable shorter term objectives. Typically shoreline management objectives include

(*a*) implementation of a strategy which ensures that appropriate measures are in place to protect against identified hazards

(*b*) management plans that are based on an adequate understanding of the dominant coastal processes and how they relate to both regional and local needs

(*c*) provision of protection schemes that provide sound economic investment

(*d*) enhancement of landscape, amenity and conservation whenever possible

(*e*) close working relationships with organisations responsible for shoreline management in neighbouring regions and organisations responsible for other aspects of "development" and "protection" within the coastal zone (i.e. an integration of the various coastal functions)

(*f*) promotion of public participation to develop awareness and understanding, and so obtain support for the management strategy.

Exactly how these objectives are formulated will depend on the statutory responsibilities of the organisation concerned. In the UK, for instance, the flooding and erosion aspects of shoreline management are covered by different Acts and responsibility devolves to different agencies. As a consequence, the National Rivers Authority is primarily concerned with

PLANNING IN RESPONSE TO CHANGE

flood protection; a Local Maritime Council with the need for erosion control (possibly closely linked with the development of tourist facilities); whereas an organisation like the National Trust regards conservation as the prime objective. The inherent limitation of individual management objectives is that they focus on the main interests of the particular organisation and do not necessarily provide for regional integration. To some extent objectives (e) above redresses this problem and notably the coastal cell groups, which have been estsblished in the UK, seek to develop a more unified approach.

Policy guidelines

Whilst sea defences are only ever provided to meet a human need, it is essential that the planning of hazard protection takes due account of other coastal interests. Having determined a need for protection at a particular location, it is necessary to establish the extent and standard to be adopted as a precursor to deciding the type and form. Equally, the factors which influence the extent and standard need to be considered on a regional basis if protection is to be provided on a consistent basis and available funding used to best effect.

This leads to the concept of Policy Guidelines. These seek to provide guidance on what policy options should be considered at a particular location and are developed using a set of rules applied on a regional basis. As such the Policy Guidelines aim to provide a framework for reviewing the standard of service provided and balancing this against protection requirements.

In doing this there are two principal considerations: the location of the defence and the standard of the defence. The first defines where, relative to the "shoreline" and the potential hazard zone, the protection could be sited, together with the physical extent of protection required. The second defines

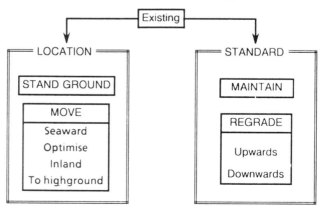

Fig. 4. Policy guideline options

the degree of protection to be provided. These two criteria are not independent and taken together provide a comprehensive coverage of the policy options available (Fig. 4). These options relate not only to new protection measures but also to the re-evaluation of existing ones. As a consequence some of the options relate to changes from the existing provisions. For instance under move location, the option to "Optimise" includes the case where a flood defence meanders and straightening out the line of the defence would reduce its overall length.

When examined regionally the Policy Guidelines can be used to develop a coherent and rationalised approach to the provision of hazard protection. The optimum combination will, however, be established on the basis of site specific studies. For instance there may be a choice between maintaining the existing line of flood defence to a standard at, or below, the target standard or moving inland and providing a defence which meets the target standard. The final choice will depend on the cost and benefits associated with each option, coupled with local preference and the ability to obtain funding.

Management response

Considering the problem of shoreline management without reference to the legislative and institutional framework, the range of response options includes

(a) *Regulation*: control of certain coastal activities (e.g. beach sand extraction, dredging, effluent discharge); modify building codes to reduce likelihood of damage; introduce insurance schemes which are consistent with strategic objectives

(b) *Planning*: land-use plans; zoning of coastal activities; subsidised relocation; emergency response planning (in conjunction with early warning systems)

(c) *Education*: make the public more aware of the dynamic nature of the coast, the attendant risks and the range of management options available; encourage participation in caring for the coast

(d) *Engineering*: reinstate the existing form of protection where this is performing well; modify the existing protection to either improve performance or reduce any negative effects; install new structures which can work within the prevailing regime

(e) *System management*: protect and rejuvenate natural habitats (e.g. saltings regeneration, cliff and dune stabilisation); remove or control cause of degradation (e.g. reduce pollution, limit access); accommodate natural movement of landforms (e.g. roll-back of barrier beach/marsh systems).

This range of response options provides for a highly flexible approach to shoreline management, but requires extensive consultation procedures to ensure that the many interests on the coast are properly accounted for. Furthermore it presumes a unified approach to shoreline management with the relevant authorities having access to at least some of the response options in each of the five main categories. Again taking the UK by way of an example, the existing legislative and institutional constraints mean that shoreline management response options are largely limited to engineering and system management.

Discussion

Having defined shoreline management as a component of coastal management and set out an appropriate management framework it is appropriate to re-consider whether or not the concept of shoreline management should be separated out in this manner. Certainly if we are to adopt a more sustainable approach to coastal management, then this too must shift from being led by resource allocation issues and move to a process based framework. As such much of the proposed management framework is equally applicable to coastal management. The core information management system would of course have to provide for a more extensive data coverage and the management strategy would need to be modified to embrace resource allocation issues (but as a management response rather than an a priori provision).

Equally, as with many other aspects of coastal management, the practice of shoreline management requires a unification of legislation and institutional arrangements, to enable a more holistic approach. It also requires access to a wide range of implements to allow appropriate responses to be developed. These arguments are consistent with the parallel case often presented to justify the adoption of an integrated legal and institutional framework for coastal management (refs 1 and 2). In essence this reflects the need for any management framework to recognise the coastal zone or region as a dynamic management entity.

The main distinction between the two is that coastal management provides the wider view and determines what activities should happen where. By contrast shoreline management ensures that an appropriate level of protection is afforded to these activities.

Given that protection is normally being provided for people and property it is essential that this is not subsumed under the wider issues of resource allocation, simply because lives are at stake. Within the UK, events such as the 1953 flood and Towyn are stark reminders of the ever present need for vigilance. This task must therefore have a strong focus.

The above arguments suggest that shoreline and coastal management have much common ground, with the potential, within an appropriate framework, for close integration. The need for a clear definition of role within a given organisation does, however, seem sufficient to justify the distinction, particularly given that human lives are at stake. One possible framework is oultined in Fig. 5. The key features of this integrated framework are that

(a) it centres on the coastal region (ref. 5); as such the strategic planning input for a given region may be from several countries (e.g. in the case of coastal regions such as the North Sea, Irish Sea)

(b) the requirements at the national level are essentially to put in place the necessary management structure at regional and local levels and to provide some strategic control over resource use as a whole

(c) the coastal classification and the associated understanding of coastal processes is provided by a management system operated at a regional level

(d) there is no requirement for each user group to duplicate data collection and analysis efforts as this is done as a joint activity

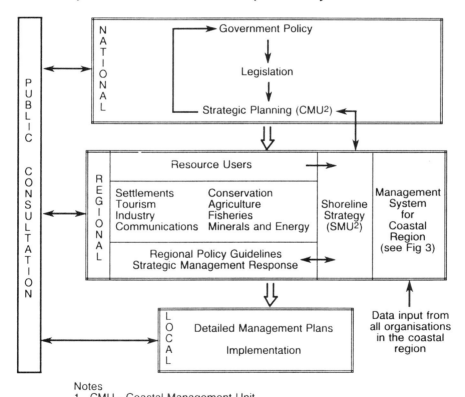

Notes
1. CMU - Coastal Management Unit
2. SMU - Shoreline Management Unit

Fig. 5. Suggested framework for integrated management

(e) provision of hazard protection is promoted by individual or groups of resource users; the task of reconciling proposals with the natural regime is done by the Shoreline Management Unit; the resultant strategy for shoreline control can then feed back into national strategic planning or directly into regional and local planning; this has the key benefit that a comprehensive range of management response options can be used to maximum effect

(f) in many situations this framework can be implemented by adjusting rather than overturning the existing arrangements; as outlined in Fig. 5 the strategic control is provided by two policy units: the Coastal Management Unit operates at a national level to co-ordinate overall resource allocation and the Shoreline Management Units operate regionally to develop understanding of the natural regime, and appropriate hazard and protection measures

(g) any conflict of interest, that might occur within a single all-embracing coastal organisation, is avoided as the roles of deriving benefit from resource use and providing protection from hazard are clearly delineated.

The exact *modus operandi* will depend on the particular legal and institutional framework for the various resource uses (degree of overlap, spatial coverage, etc) and the physical nature of a particular coastal region. Clearly a region comprised largely of rocky shores, rural communities and localised pressure for recreation and tourism is going to have very different needs from a region that is low lying, subject to flooding and erosion, with heavy urbanisation.

By starting with a proper understanding of the natural environment as the basis of planning and management it should be possible for the many and varied coastal regions to be managed from within the sort of coherent management framework outlined above.

References

1. Gubbay S. 1989. Coastal and sea use management: a review of approaches and techniques. Marine Conservation Society.
2. Townend I.H. 1992. Coastal management. In: Coastal, estuarial and harbour engineers reference book. Chapman and Hall.
3. Fischer D.W. 1985. Shoreline erosion: a management framework. Jnl Shoreline Mgmt, 1, 1, 37-50.
4. Townend I.H. 1990. Frameworks for shoreline management. PIANC Bulletin, No. 71, 72-80.
5. Vallega A. 1985. Coastal planning and regional coastal planning: the seaward patterns of spetial organisation. International Geographic Union Proceedings, Geneva.

6. Salm R.V. and Clark J.R. 1984. Marine and coastal protected areas: a guide for planners and managers. International Union for Conservation of Nature, Switzerland.
7. Townend I.H. and Fleming C.A. 1990. The Anglian Sea Defence Management Study. Proceedings of Flood Plain Management Conference, Ontario.
8. National Rivers Authority. 1991. The future of shoreline management. Conference Proceedings, NRA, Anglian Region.
9. Barber P.C. and Townend I.H. 1991. Monitoring guidelines. Report for the National Rivers Authority, Anglian Region.
10. McCue J.W. 1989. The use and potential use of GIS in coastal zone management. MSc thesis, Department of Biology, University of Newcastle-upon-Tyne.

Glossary

Coastal characteristic - a distinguishing feature of the coast, either natural or human (e.g. morphology, ecology, usage)

Coastal forcing - the natural processes acting on the coast (e.g. winds, waves, tides)

Coastal processes - the interaction of coastal forcing with coastal characteristics (e.g. sediment transport, pollutant dispersion)

Coastal response - change resulting from coastal processes (e.g. erosion, flooding)

Coastal resources - the coastal characteristics which can be used for human benefit (e.g. life support, economic gain, recreation)

Coastal elements - any physical or human component on the coast, this being the aggregate of characteristics, forcing and response

Shoreline - one characteristic of the coast; poorly defined but essentially the interface between land and sea

Shoreline management - relates to efforts to control the shoreline interface; as such, this usually focuses on coastal hazards and the need to provide protection against either flooding or erosion

Coastal zone - some combination of land and sea area, delimited by taking account of one or more elements

Coastal zone management - the planning and allocation of resources within a given area

Coastal region - a spatial region where the elements have a degree of independence from external elements and the potential to be organised within an integrated framework

Coastal region management - the planning and implementation of a structure plan, embracing all elements, for the development of a coherent spatial region

PLANNING IN RESPONSE TO CHANGE

Management framework - the complete structure, which includes all aspects of the management process (i.e. data inputs, analysis and understanding, planning and allocation, dissemination and consultation, regulation and enforcement, etc)

Management system - the tools used to develop and implement a management strategy; these can range from paper based archives to a computer database with facilities to present information spatially and undertake various forms of analysis

Management strategy - sets out the requirements for a management plan; the aims and objectives set out what is to be achieved and will be judged against defined criteria; implementation will make use of policy options or guidelines and an appropriate range of response options

Management plan - defines how strategy will actually be implemented

Management aims or goals - focus on long term aims and may be somewhat open ended

Management objectives - shorter term aims which are measureable and provide a means of fulfilling the goals

Asessment criteria - the measures or standards used to develop the management plan and subsequently assess progress towards implementation

Policy options/guidelines - a definition of resource use and constraints (e.g. promotion of tourism development in an area, limits on mineral extraction, protection of a particular habitat, location and standard of sea defences)

Management response options - means by which policy options can be implemented (i.e. regulation, planning, education, eco-management, engineering, etc)

Legal framework - defined by Government policy and ensuing statutes

Institutional/Organisational framework - devolve from the legal framework as the basis for implementing the statutes

Conservation - the protection of an area, or particular element within an area, whilst accepting the dynamic nature of the environment and therefore allowing change

Preservation - static protection of an area or element, attempting to perpetuate the very existence of a given 'state'

18. Coastal zone planning beyond the year 2000

I. R. WHITTLE, MICE, FIWEM, MBIM, formerly Head, Flood Defence, National Rivers Authority, now retired

Within an undefined area of land in England and Wales, referred to loosely as the coastal zone, the National Rivers Authority (NRA) is charged with the duty of exercising general supervision over sea defences, and specific duties and powers in relation to water quality, and aspects of sea fisheries, recreation and conservation. The Authority is a consultee under planning legislation. However, the Authority is only one of a plethora of bodies which operate in the coastal zone largely in an unco-ordinated manner, even though the activities of one may have impact on another. There is much pressure from institutions, public bodies and individuals and in certain government circles for the establishment of a central organisation. Such a body would be responsible for formulating policies and be charged with the task of developing and implementing an integrated strategy so that all the coastal zone resources are utilised in a sustainable way in the years to come.

This paper examines the possibility of the NRA fulfilling that role in the next century. It is set against the background of a series of formal inquiries. The House of Commons Select Committee for the Environment is holding an inquiry into Coastal Zone Protection and Planning; the National Audit Office is investigating coastal engineering scheme administration in MAFF, the NRA and local authorities, and the government announced in 1991 its intention of creating a new environment protection agency. Whatever the outcome of these inquiries, and whichever option is decided for an environmental agency, there is likely to be a significant change in the structure and role of the NRA. Although this paper examines a possible role for the NRA, it may be the government's decision to establish a new agency for the protection of the coastal zone in the next century. It then remains a question whether or not that agency is given any wider responsibility for flood defence matters.

Introduction

The NRA was vested as a New Departmental Public Body in September 1989 under the Water Act 1989. That act transferred responsibility to the NRA for all the functions of the former Regional Water Authorities, with the exception of water supply and sewerage and sewage treatment which are the responsibility of the Water Authority PLCs.

PLANNING IN RESPONSE TO CHANGE

Within the coastal zone, the NRA has a range of powers, responsibilities and duties, many complementing those carried out by the numerous organisations which have interests in the same geographic area.

Two of the principal functions performed by the NRA in fulfilment of its flood defence duties in the coastal zone are the construction of appropriate sea and tidal defences and making recommendations to local planning authorities and councils, in response to either planning applications or in the preparation of development and structure plans.

However, whilst these functions impact upon or are impacted upon by many of the activities of those other organisations (Fig. 1), the Authority does not have any statutory safeguards in planning decisions within flood risk areas or effective control over other activities which affect the coastal processes.

Control of activities in the coastal zone is subject to much control by both public and private legislation. Acts cover planning control, construction of new sea defences, erection of coast protection works, construction of marinas and similar activities, for which the NRA is a statutory consultee. Through this administrative provision the NRA interests are assumed to be safe-

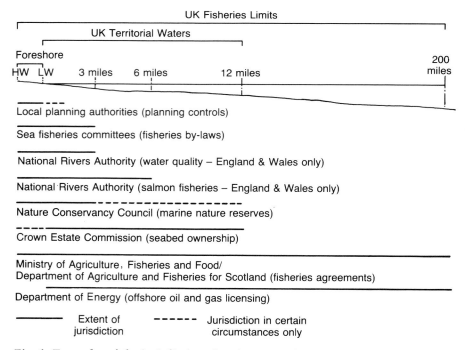

Fig. 1. Examples of the jurisdiction of various organisations in the coastal zone on specific issues (after Gubbay, ref. 1)

guarded. This is not necessarily so in practice because the NRA is totally dependent upon the co-operation and support of the planning authorities. There have been occasions when the NRA's advice or recommendations have been ignored, with predictable consequences. Although rarely tragic, these occasions have resulted in much misery, financial hardship and long-term human stress.

Historically, the extensive range of engineering works of port developments, coast protection, sand and shingle extraction from the shoreline and the construction of sea defences, together with natural coastal processes, have proved deleterious to some sea defence structures.

Although the NRA is consulted by Maritime councils over proposed works to protect against erosion, the Authority has no jurisdiction to require them to undertake such works.

Over time, there has been ever increasing evidence of damage from the inter-action of coastal processes and structures. Only recently have engineers and scientists, using numerical and physical modelling, been able to understand better these complex processes and assist with the predictions of changes to the shore line and the possible impacts upon coastal defences.

The stability of coastal defence structures, cliffs and foreshore is dependent upon an adequate supply of beach-building material. Any interuption to that supply, by say the construction of a breakwater or pier at a harbour entrance or by the removal of material from the rear shore area, will adversely affect that often fragile stability. Whilst the removal of such material from above the high water mark (HWM) is subject to control by the local planning authority, this is not so below the low water mark (LWM). This latter activity is controlled by the Crown Estate Commissioners (CEC).

The CEC is responsible for issuing licences for the extraction of minerals below the LWM but they are constrained so that materials can only be 'won' from water depths greater than 18m. Coastal experts consider that removal of material at this depth does not pose a threat to the quality or the stability of the foreshore or adjacent defences.

The legislation setting up the NRA retained provisions for the establishment of Regional Flood Defence Commitees chaired by a MAFF appointee, and it required the NRA to delegate to those committees responsibility for all flood defence functions except the issuing of levies.

Each committee is constituted so that the majority is held by representatives from those local councils responsible for raising the flood defence funds (called levies) required by the NRA.

Council members are required to ensure that their council's overall plans are not jeopardised. A balanced approach is taken when setting levies, and a conrtol mechanism operates above a limit of 2.5 times the penny rate product calculated from the former 'rateable value' of property, updated to allow for inflation. These Council members are required to pass a special resolution to exceed this limit before a flood defence committee can approve an annual expenditure plan, and through this financial control the planning and funding of sea defence activities appears to remain at local level. This control exercised by this balance in the membership is generally regarded as appropriate in a democratic society, but this does mean that occasionally high priority works may be deferred.

Overall this executive structure has proved successful and it has promoted the construction and maintenence of robust defences which have failed rarely and then only when subjected to forces greater than the design limits.

In reality the expenditure on sea defence by the NRA does not fall upon the local population. The Community Charge process administered by the Department of the Environment (DoE) results in a reimbursement of the flood defence levy to the relevant councils through Revenue Support Grant made under the Standard Spending Assessment (SSA), although the levy has to be set against the council's spending and counts toward rate-capping expenditure. Thus, it is clear that central government already funds the total expenditure on flood defence. However, because of the present legislation it does mean that responsibility for flood defence planning does not lie with the Board of the nation's principal defence authority, i.e. the NRA. The Board is required to issue levies and in doing so has regard to needs set out in the Aurhority's Corporate Plan. The annual planning process involves discussions with two government departments (DoE and MAFF) and so the Authority can highlight areas of concern through this consultation procedure.

This funding and planning arrangement is complex and whilst it does provide a focus for parochial interests, it must be questionable whether or not it provides a wholly cost-effective result, giving value for money. There is a view expressed that a rationalisation of this present arrangement would improve the business efficiencey of both the NRA and MAFF.

The NRA is required to further the conservation and enhancement of the environment and the unique nature of the coastal zone provides many opportunities to fulfil this duty. The NRA is not alone in this: it is imposed also upon many other authorities and institutions working in the coastal zone. Guidelines are currently being prepared to assist the NRA and other organisations on ways to protect and enchance the coastal environment.

Towards the year 2000

Over the centuries, man's activities, together with natural coastal processes, have impacted upon the total coastal environment and have threatened and affected this fragile area. There is now the further threat from a predicted rise in sea levels due to global warming. In a geological timescale sea levels have fluctuated more than currently predicted. This new threat has been one of the factors behind the renewal of interest in the coastal zone.

Other factors enhancing this interest arise out of Directives emanating from the European Council, implemented by the ensuing legislation, and pressure from British environmental agencies drawing attention to shortcomings in protection policies. These organisations are pressing for integrated policies for safeguarding and sustaining the coastal zone. Strategic policies can only be developed when they are soundly based upon a detailed knowledge and understanding of all processes. Many of the complex inter-actions are capable of being modelled and the modern computer capability facilitates long term prediction.

Lately there have been numerous new initiatives to study the many varied features of the coastal zone. The NRA is undertaking studies to develop management strategies for protecting low-lying lands. In particular, a major study has been undertaken in the Anglian region.

This Sea Defence Management Study (SDMS) examined a range of coastal processes and gathered data about the natural forcing mechanisms. It recorded the evolution of numerous features, many of which are inter-dependent, and rely upon the complex processes found in the natural environment. The subsequent evaluation of this comprehensive data set has produced a series of possible defence options.

These options range from the traditional solutions involving 'hard' engineering through to the 'soft' engineering, such as beach recharge. Other options include 'do-nothing' or, in appropriate cases, 'retreat' to form a new defence line. A policy which includes these latter options would find much favour with many agencies, but use on a wide scale would mean reviews by many agencies and a change in the legislation.

The recent national Sea Defence Survey (SDS), undertaken by the NRA, has provided, inter alia, a database of the locations of small, low-grade areas of land. The continued protection of these areas may not be cost-effective and, especially, might not serve the best interests of the environment.

A report prepared jointly for the NRA and other bodies (ref. 2) examined the possiblity of using these areas for recreational activities and environmental enhancement. It suggests a range of alternative strategies for the existing defences ranging from 'do nothing' to 'managed retreat', controlled reduction in defence standards.

The construction, reconstruction or maintenance of a flood defence may not be undertaken when the costs exceed the value, in terms of output or damage-avoided, of that protection. For this reason there exist areas where defences have been abandoned, and this practice may become more common if sea levels rise. Often the environmental or amenity value of these areas is low, but with sound strategic planning the value may be enhanced to a level to justify investment of public funds.

PLANNING IN RESPONSE TO CHANGE

Beyond the year 2000

In 1991, a number of initiatives were announced which might lead to changes in the administration controls in the coastal zone.

The government announced its intention to establish a stronger, wider-ranging and more powerful environment protection agency. The exact form of this agency has not yet been determined nor, therefore, has the future of the NRA been confirmed. There has been the suggestion that the responsibility for flood defence matters might be removed from a re-structured NRA. A change may also stem from the outcome of the House of Commons Environment Committee inquiry into Coastal Zone Protection and Planning carried out in early 1991. Some of the evidence received by the Committee will have recommended almost certainly the creation of a Coastal Agency. Such an agency, perhaps structured on the lines of the Broads Authority which crosses administrative boundaries, would need wide-ranging executive powers to ensure integrated development of this important area of the UK.

This Authority could possibly be given some form of planning veto, a power which would not be appropriate for the NRA in its present form. An alternative could be a strengthening of the role of the coastal cell groups, if necessary supported by appropriate legislation. These groups have expanded recently, both in size and number, so that most of the vulnerable lengths of coastline are now represented. The framework of the groups is already established but the constitution would need to be rationalised.

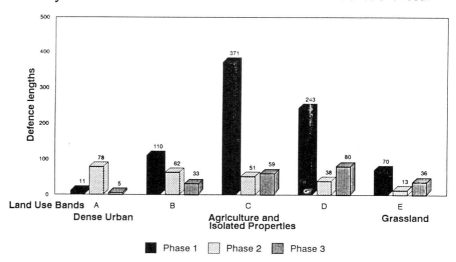

Fig. 2. National Sea Defence Survey: lengths of defences protecting land use bands, phases 1 (NRA defences), 2 (local authority defences) and 3 (institutional and private defences) (courtesy NRA)

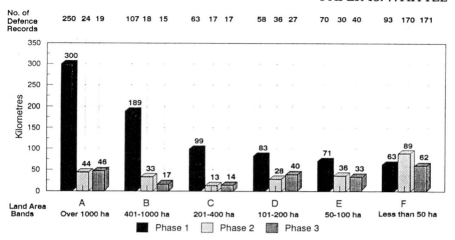

Fig. 3. *National Sea Defence Survey: lengths of defences protecting land area bands, phases 1, 2 and 3 (courtesy NRA)*

Another alternative exists within the Coast Protection Act where there is a provision for the creation of a coast protection board (ref. 3): a provision which has never been implemented. Whatever institutional arrangement is eventually decided upon that institution of defence agency will need to address the issue of sea level change as one of its top priorities.

Until recently, scientists were predicting an acceleration in the rate of global warming with a consequential rise in sea level. The scientific uncertainties are exemplified by constant revisions to the predicted rate of rise and the current predicitions are for a lower rate of rise. Whatever the rate, the implications of a rise will lead to changes to the coastal environment.

The rise in sea levels put at a greater risk the whole regional economy and infrastructure. Future scenarios have been promulgated and the possible effects on the socio-economic state of regions has been assessed by government departments in direct cost terms. The value of possible change to the environment has not been evaluated because no suitable financial model is available. If the environment is altered significantly by the predicted global change, such a model could affect future defence strategies.

It is certain that the major urban areas will continue to be protected, but the strategies for the smaller, lower-value lands will need to be reviewed. The 'retreat' or the 'do-nothing' options will probably be adopted more extensively than at present.

The large areas of agricultural land are fairly certain to be protected. Some 57% of the 348000 ha of Grade 1 lands (MAFF classification) (ref. 4) lies below the 5 m AOD (Newlyn) contour. All of this land is currently defended by schemes designed against the 'worst recorded storm tide' (ref. 5). The SDS has shown that some 29% of the NRA's sea defences (of 805 km) is fronted by an eroding foreshore. This problem can only get worse if the climate change brings increased storminess to cause damage to those foreshores.

Figures 2 and 3 show the extent of exisiting sea defence problems which will need to be addressed. These and the related complex problems of integrated coastal management will have to be solved through closely-meshed strategy groups, working through a single co-ordinating body.

Conclusions

If the coastal zone is to survive in approximately the form in which it presently exists then improved coastal planning will be necessary. The damaging effect of past actitivies, whether man made or natural, require re-assessment to devise mitigation programmes. The many pressures on this zone must be co-ordinated so that each can be complementary.

This paper has suggested some alternative administration frameworks to co-ordinate the various pressures. Ideally, the selected administrative body should be independent of many constraints placed upon either local authorities or pressure groups. At the same time local input is necessary to provide a sense of direction and 'ownership'. This is best achieved by a national body working though a small number of regional units and this sample framework is easily visualised. The NRA, if given greater powers, is ideally placed to fulfil the role of the co-ordinating body, should the outcome of the present series of inquiries reject the principle of a special coastal agency or any of the aforementioned options.

Acknowledgement

The Author is grateful to the NRA for permission to present this paper. The views expressed in this paper are those of the Author and may not necessarily represent those of the National Rivers Authority.

References

1. Gubbay S. A future for the coast? A report for the World Wide Fund for Nature, from the Marine Conservation Society, 1990.
2. Posford Duviver Environment. Environmental opportunities in low lying coastal areas under a scenario of climate change. National Rivers Authority, Department of Environment, English Nature (Nature Conservancy Council), Countryside Commission, 1991.
3. Coast Protection Act 1949, 5.2(1). HMSO.
4. Whittle I.R. Lands at risk. Loughborough Conference of River & Coastal Engineers, MAFF, 1989.
5. Flood Protection Research Committee. MAFF, 1979.

Discussion

While the need for a lead authority in Coastal Zone Management (CZM) was emphasised and was the main focus of the discussion, it remained unclear which body should take that lead. The complexity of the problem was acknowledged and it was recognised that no single body could be charged with the execution of all the relevant functions, but rather it was accepted that the lead authority should give the framework and guidance within which existing bodies could function in a co-operative manner.

The NRA was suggested as a possible lead authority but problems were raised due to the NRA not being directly accountable to the public and currently not having any planning functions. In general it was observed that if significant responsibility were retained by local government for CZM it would be difficult to make them answerable to a non-elected body.

The coastal groups were cited as a possible mechanism to drive CZM as they had been shown to consult and work. Others felt that they were too fragile, unstable and parochial. Furthermore, the various coastal groups were differently constituted and if they were to carry a national role would need to be reconstituted to a common pattern. Similarly the Broads Authority and National Parks were suggested as suitable models but with no decisive acclaim.

Interest in coastal management had moved significantly forward since 1989 and the question was asked as to how the co-operation which was apparent at the Conference could be kept going. Was a working party a possibility? Comments were made that the government attitude had been complacent and that the Conference was an opportunity to put forward views for consideration in conjunction with any Government review arising from the Select Committee report.

19. Beaches - the natural way to coastal defence

DR A. H. BRAMPTON, Coastal Group, HR Wallingford

Introduction

Development of the coastal strip for industry, housing and leisure has led to the construction of a wide variety of solid structures such as promenades, reclamation walls, and road/railway embankments. Many of these structures now also serve as defences against erosion or flooding. In most cases, a beach forms a vital part of the defences, but often by default rather than by specific design.

By continually adjusting its shape, a beach absorbs much of the incident wave energy, hence reducing the forces on the structures behind. Long-term erosion will reduce the capacity of a beach to perform this function, and also allow the undermining of the defences themselves. The amenity value of the coastline is also reduced.

Recognizing the key role of beaches in protecting the coast, and acknowledging the recreational opportunities and the need for environmental conservation, engineers are increasingly turning to 'soft' defence options. Active beach management techniques such as periodic nourishment, by-passing and re-cycling are now commonplace alternatives to strengthening or rebuilding seawalls. There are now a sufficient number of such schemes operating in the UK to demonstrate that they can be safe and economically worthwhile. However, designing a beach to provide a specified level of protection over a required lifespan is still a major problem. This paper concentrates on the role of beaches, whether natural or man-made, in defending the coast. In particular, attention is focused on methods for predicting beach morphological changes, both well-established and novel.

Natural beaches as role models

If existing, solid defences are to be protected by a 'managed' beach, the logical way to proceed is to base its design on existing local beaches. Indeed there is considerable pressure to choose just this strategy. There are clear advantages from both aesthetic and ecological viewpoints in adopting the equivalent to a vernacular style of architecture. It is also likely that there will be an economic advantage in using locally available materials.

It has to be recognized, however, that a healthy natural beach that has endured wave attack for centuries does not necessarily provide a satisfactory coastal defence. Massive barriers such as Chesil Beach have been overtopped - even breached - in the past, with subsequent serious flooding of the hinterland. With so much development close to the coast in modern times, similar failures in the future could be catastrophic. Artificially built or strengthened beaches will therefore have to be more substantial than their naturally created neighbours.

Showing that even a substantial beach will be an adequate defence is no easy matter. As in many other walks of life, there is an ever increasing requirement for certainty, especially when there is a threat to life or property. Solid coastal defences can fail suddenly, and without warning, leading to serious flooding or erosion. Despite this, the populace have grown accustomed to massive seawalls, and regard artificial beaches with suspicion, sometimes hostility. It is important to demonstrate the safety of a beach under severe conditions, and also to provide a well justified estimate of its likely maintenance requirements and its expected life. This presents a challenging task when dealing with beaches and their responses in the complex and hostile coastal environment.

It is necessary, therefore, to use predictive techniques, principally numerical simulations or scale physical models, to examine the behaviour of beaches. In extreme conditions, for example storm waves occurring at the same time as an exceptional tidal level, it is the short-term response of a beach profile which is of prime concern. In the long-term, however, it is likely to be the movement of beach material along the coast which requires more attention.

Understanding the morphological changes of natural beaches, and particularly the causes of erosion, is vital. Because of this, measurement of waves, tides, and especially beach changes is an essential component of the 'soft' defence strategy. The data obtained not only helps to demonstrate the performance of any scheme, but also provides both calibration of, and inspiration for, the methods used to predict beach behaviour.

Numerical modelling of alongshore sediment transport

It is convenient to start a discussion of the various predictive methods used to study or design beaches by separating the movement of beach sediments into two components, namely along the coast (i.e. parallel to the beach contours), and cross-shore (i.e. perpendicularly to the contours). As will be mentioned later, this is a simplification which can now be avoided, but will serve for the time being.

In most situations where beach erosion is rapid, and persists over a period of months or years, the cause is a change in the pattern of alongshore transport of sediment. One eminent coastal engineer (ref. 1) recently stated that

"... all examples of shore erosion on nonsubsiding sandy coasts are traceable to manmade or natural interruptions of longshore sediment transport"

This is certainly an over-simplification but serves to emphasise the importance of alongshore drift in the evolution of beaches, both in the UK and overseas. When material has been lost from beaches that have been nourished, for example, it is usually this mechanism that has caused the damage.

Against this background it is not surprising that much of the effort exerted in predicting beach evolution has been spent in developing models of alongshore sediment transport and the changes in beach plan shape that it can cause. Simple computer methods that can, for example, simulate beach accretion and erosion on opposite sides of a harbour or long groyne have been in routine use for over 20 years, and are now available as off-the-shelf packages from a variety of specialist consultants around the world.

The most common type of model is the so-called "one-line" method in which the advance or retreat of the beach face is deduced from predictions of the changing position of a single, representative contour (usually the high-water mark or mean sea-level). By definition, such models do not predict changes in beach profile shape but if the objective is to simulate the long-term changes in a beach, over several years or more, then this is not a serious drawback. Net changes in the beach profile over such a period are usually small compared with those brought about by alongshore sediment transport.

The formulations of the equations used to calculate the rate of transport are generally rather simple, and often partly empirical. As a result they may not be sufficient to predict accurately the movement of beach material over a few hours or days in a severe storm. Despite this, long-term modelling has proved to be a powerful technique, especially on relatively straight coasts when there is good historical information which can be used for calibration. The simplicity of the one-line model has also been an advantage in tackling more complex coastline problems. The original validation of this type of model was for the long, straight, sandy beaches of California (ref. 2) where tidal currents were low and waves were often near-monochromatic in character (i.e. of single wave height and period). Over the years, however, very similar models have been developed to study the changes in beaches along the coast of the UK, where tidal ranges are high and the beach material is often of shingle (ref. 3) or a mixture of sand and shingle.

Recent developments have included the ability to model the effects of groynes on shingle beaches, a necessary improvement given the large number of such structures around our coasts. Since such groynes are rarely long enough to prevent some transport of shingle past their tips, it has been necessary to calculate the variation in the alongshore drift rate across the beach profile. This involves using information on the level of the tide at any instant, expressed in probabilistic terms, together with an expression for the distribution of wave-induced transport of beach material at different depths relative to the tidal level. At prssent the distribution function is derived from experimental data, but could equally well be calculated from numerical modelling of processes acting over the whole beach face. This extra element in the modelling of alongshore transport has identified an interesting feature affecting coastal development. At several sites around the coast of the UK, it has been found that the net direction of alongshore sediment transport is different at different beach levels. Small but persistent wave action may produce a drift northward on the upper part of the beach profile, for example, while occasional large storm waves can cause a southward movement lower down the beach face.

This effect may also be caused by tidal currents, which modify the wave-induced drift of beach material, particularly on the seaward portion of the active beach profile. Again recent modelling of beaches in the UK has suggested that the situation where drift is in opposite directions at the beach crest and the toe of the beach profile is more than a theoretical oddity.

The modifying effect of tidal flows on beach sediment movement greatly complicates any modelling exercise. It is difficult enough to produce sufficient, accurate wave data (recorded or hindcast) to drive a model of beach evolution. The increase in effort to include the tidal currents and variation in water levels would be a serious additional burden. In view of this, considerable effort has been spent in 'input filtering', that is to say in reducing the amount ot data needed to predict successfully the changes which a beach will undergo. One simple example is ignoring all waves with height less than a certain threshold value, because they would not have sufficient energy to move the beach material. A more complicated idea is to adjust the formula for wave-induced sediment transport to reflect an 'averaged' effect of tidal currents. This involves the use of a subsidiary model, run prior to the main beach model, which calculates the total effect of both waves and tides on alongshore transport rates over a complete tidal cycle (or even over a Spring-Neap cycle) for a wide variety of wave conditions. This results in a more complex model, but one which can still be run relatively easily, for long periods, and with sufficient resolution to examine alternative groyne layouts, for example.

The major requirement for such models is now good calibration data, from laboratory experiments but more importantly from suitable coastal sites. The morphological data from the coast will need to be accompanied by information on wave conditions as well. Numerical experiments, supported in part by historical data on coastline changes, have shown very large variability from year to year in the alongshore transport at many sites around the UK coast. This fluctuation in net annual drift rates makes it very difficult to define an average value.

Typically, it may take 10-20 years to produce a statistically reliable estimate of the mean value, by which time climate changes may well be a factor needing to be taken into account.

Numerical modelling of cross-shore sediment transport

The variability of beach levels, particularly against the face of a seawall is a source of considerable excitement and concern around the coast of the UK, and the world outside. Local residents - and sometimes the local coastal managers - are often convinced that their own stretch of coast is unique in this respect, and talk in awed tones of the sudden fall in levels during a single tide, i.e. between successive low-water periods. Levels against a defence may fall even lower at the time of high water than revealed by a survey a few hours later when the beach dries. A method of recording beach levels continuously during a storm would provide valuable, new information either for natural beaches or those with defences at their rear.

On an open, unobstructed beach (i.e. without groynes), short-term morphological changes will be dominated by cross-shore transport mechanisms. Despite this, modelling the movement of sediment up and down the beach face has been neglected in comparison with alongshore drift studies. In the UK particularly, the large tidal range usually prevents the formation of significant nearshore 'bars' which intrigue coastal research workers in many other countries.

Models currently available to predict beach profile changes fall into two categories. On one hand there are empirical techniques, usually based on relationships between the beach profile shape and the incident wave parameters. These models are based on analysis of laboratory experiments, and make no attempt to represent the transport mechanisms. They are, however, easy to apply and cheap to use. Limited field measurements also show that the laboratory experiments give realistic representations of beach profile response to wave action, particularly when dealing with shingle.

The alternative approach, being developed in many countries around the world, is first to model the hydrodynamics affecting the beach, and then to calculate the resulting transport of sediment. This is clearly a more satisfying approach from a scientific viewpoint, but some degree of approximation is still required to obtain reasonable results. The models of this type also allow the time development of a profile to be studied, rather than presenting just the final 'equilibrium' profile for any wave condition. Understandably, such techniques require much greater computational effort and have been slow to develop. The most recent versions are beginning to show encouraging signs of reproducing beach profile changes measured in large laboratory experiments.

Even so, a recent paper on beach response to storm waves (ref. 4) states that the predictive skill of such models is very low. This is hardly an encouraging assessment for those needing to demonstrate the safety of artificial beaches in extreme conditions. Little convincing work has yet been done on the effects of man-made structures on the development of beach profiles using such models, nor on the interchange of sand between dunes and the beach face.

It is also becoming clear that wind waves alone do not necessarily determine the changes in beach profiles, or the associated cross-shore transport of sediment. On oceanic shorelines especially, the primary (i.e. wind-generated) waves produce underlying 'long-waves' which have periods of a few minutes. These long-waves are much more important than the primary waves when considering certain aspects of coastal engineering, particularly the movement of large vessels within harbours. A number of research studies point to the possibility of such long-waves being important in moving sediment on the face of a beach, and thus being a factor in shaping the beach profile. If this is indeed the case, there will need to be considerable further work on wave prediction in shallow water to produce appropriate input conditions for beach profile models.

Before leaving the subject of beach profile models, a further difficulty needs to be highlighted. The models of beach profile development are usually only used for a short time period, typically long enough to consider the effects of a single storm event. It seems likely that 'upscaling' these models to predict profile evolution over weeks, or months, would be very difficult, and perhaps impossible, at least in the foreseeable future. There is no equivalent, apparently, to the long-term one-line models used for beach plan shape predictions, which can be 'downscaled' to predict short-term changes without exciting great concern.

The problem of long-term beach profile prediction is further complicated by gradual effects not included in the normal models. These include sea level change relative to the land, erosion of the often soft rock substrate on which the beach material rests, and losses of material both seaward and landward. These losses may include wind-blown sand moving landward of the active dune system, or the dispersal offshore of fine-grained material, as well as direct human intevention (e.g. sand-mining).

The only method of predicting long-term beach profile changes at present therefore appears to be to statistically analyse historical data, for example regularly surveyed beach cross-sections, and to extrapolate into the future using the trends identified. This further strengthens the case for monitoring beaches by coastal authorities charged with their management.

Numerical coastal area modelling

The previous two sections of this paper have discussed the modelling of beach morphology by separating the sediment transport into perpendicular components, along the contours and parallel to them. Interactions between the transports in the two directions are either ignored or only taken into account in a simplified way. Recent developments in the modelling of coastal changes have allowed this simplification to be overcome.

First, for straight or nearly straight beaches, it is possible to combine the methods already developed for the alongshore and cross-shore sediment transports, and so produce a model which can calculate, at least in the short-term, the changes in morphology in a single operation. One such model was developed, and subsequently employed as part of the NRA (Anglian Region) study of the management of sea defences between the Humber and the Thames Estuary. It can be anticipated that such models will become a standard tool in the coastal manager's tool-kit in the future.

In situations where the sea bed contours, or the solid geomorphology of the coast is complex, however, the basic assumption that waves and tidal currents are shore-parallel and vary only slowly along the beach is untenable. In this situation it is necessary to change the basis of the beach morphological model completely. A grid system has to be established over both the beach and the nearshore sea bed, and the hydraulic forces, i.e. the waves and tidal currents, evaluated at each mesh point. This is then followed by calculation of the sediment transport vectors over the whole grid, leading to deductions on the resulting morphological changes. The whole cycle of operations is then repeated. Since such models have been developed to deal with beaches which are not of a simple shape, it can be appreciated that the computational effort involved can be enormous, even to consider changes over a few hours or a single tidal cycle. The development of these complex models has therefore required a great deal of thought to decide on both the range of processes that should be included, and the degree of detail required to represent each (process filtering), and also the number of different input conditions (e.g. tidal range, wave height) that are needed to provide useful predictions (input schematisation). The development of both these powerful coastal area models, and the methods used to apply them in practical situations, will be the subject of many further years of research worldwide.

Physical modelling of beach morphology

Given the difficulties, and cost, of using numerical models of complex beach areas, for example in the lee of a system of offshore breakwaters, scale physical models remain a major force in predicting coastal changes.

The introduction of sediment into scale models always requires careful planning, and validation from prototype beaches whenever possible. In addition, it is normally necessary to use a numerical model alongside a physical model, in order to evaluate the sediment transport rate scaling (which is generally different from that for the water velocity scale). A discussion of this interactive use of physical and numerical model techniques has recently been presented (ref. 5) which provides further details.

Given this initial effort, which is certainly required for numerical models as well, the use of scale physical models can certainly provide useful predictions of short-term beach changes. It is worth making the point that such models have also provided much of the calibration data for many of the numerical modelling methods now in everyday use.

At the same time, there are clear limitations on such models. It is difficult to model more than 1-2 km of coastline in a normal laboratory basin. Time and cost constraints also prevent the use of physical models for longer than a few weeks (prototype), so that beach evolution over longer periods has to be extrapolated using numerical techniques.

Beach monitoring

If we are to use beaches as a method of defending our coasts, the best results will be achieved by a combination of careful design with subsequent monitoring and analysis.

One consequence of the growing use of beach nourishment has been an increase in the surveying of the coast. In this situation, simple analysis of the volume of beach material provides an easily understood way of checking on the success of the defence scheme. Changes in volume can be converted into monetary terms, and the cost-effectiveness of the works can be assessed. This is much less simple to do when assessing the cost-effectiveness of other types of defence, for example groyne systems.

Monitoring beaches can achieve much more than this, however, and it is worth setting out the main advantages that could be gained from regular surveying. First, it is clear that money available for defending the coast should be allocated on the basis of demonstrable need. Surveying and analysing the coast provides evidence to support assertions that beaches are likely to fall to dangerously low levels in the future. Using the observed trends, and information on severe wave and tidal conditions, it is then possible to evaluate the likely future performance of defences, and arrange a programme of improvements well in advance. Strategic planning of this sort is not common around the coast of the UK at present, and awaiting the failure or near-failure of defences rarely allows an optimal solution to be installed.

Monitoring should also provide useful information on the relative performance of different types of defence works. As an example, it is still rare to see beach profiles compared with the cross-section of a groyne. It is difficult to see how groyne profiles can be improved without this basic information. Similarly, many authorities have a variety of sea walls along a single frontage. Measuring beach levels close to the face of defences should indicate whether different slopes, roughness, or permeability could produce a healthier, less volatile upper beach.

Finally, from a scientific viewpoint, the data obtained will be invaluable in the development of predictive models of beach behaviour. Better understanding of natural or artificial beaches can only help in the adoption of 'soft' defences which are safe, economical and environmentally sensitive.

References

1. Galvin C. (1990). Importance of longshore transport. Shore and Beach, vol. 58, no. 1.
2. Komar P. D. and Inman D. L. (1970). Longshore sand transport on beaches. Journal of Geophysical Research, vol. 75, no. 30.
3. Brampton A. H. and Motyka J. M. (1985). Modelling the plan shape of shingle beaches. Lecture notes on Coastal and Estuarine Studies, vol. 12, Springer Verlag, Berlin.
4. Huntley D. A. (1991). Beach response to waves and currents detailed field measurements. Developments in Coastal Engineering. University of Bristol.
5. Brampton A. H. et al. (1991). Application of numerical and physical modelling to beach management problems. Proc. 3rd International Conference on Coastal and Port Engineering in Developing Countries, Mombasa, Kenya, pp. 180-194.

20. Engineering the beaches

G. M. WEST, BSc, MICE, MIWEM, Flood Defence Manager,
National Rivers Authority, Southern Region

This paper reviews the history of sea defences and works that have been undertaken in the Southern Region of the National Rivers Authority. It poses the question 'Why defend?' and points out the need for improvements in coastal zone planning, especially as a consequence of sea level rise. The coastal cell should be considered as an entity and there should be full co-operation with all coast authorities, co-ordinated by MAFF. The need for project appraisal is stressed as is the fact that all options should be fully evaluated.

Introduction

The inhabitants of Great Britain have defended themselves against the sea for many centuries, back to Roman times. Nowhere was this more so than along the southern coast of England. Romney Marsh, the cradle of land drainage, was protected from the sea by the Great Seawall at Dymchurch. Roman remains have been found at Dymchurch and records exist that show that specific works were undertaken in the reign of Henry III in 1251 to defend the land from the sea.

The coastline from the Thames Estuary to the Hampshire/Dorset border is some 1000 km in length, of which 143 km are sea defences maintained by the National Rivers Authority (NRA). There are basically two types of defence: coventional seawalls - hard defences, and the more abundant in length shingle ridges - soft defences.

With the defences open to attack from the predominant south-westerly gales, there is a continuing need to maintain effective defences, especially with the advent of sea level rise, resulting from climatic change, a consequence of global warming.

The recent National Sea Defence Study undertaken by the NRA has identified that the vast proportion of sea defences in England and Wales are in a satisfactory condition. Those priority works in Southern Region are contained within the current Medium Term Plan. Specific strategic studies are underway at Dymchurch, Pevensey Bay and Elmer, Bognor Regis. In addition, further investigations are proposed for the Northern Sea Wall (Reculver to Birchington) in Kent and between Shoreham and Lancing in Sussex.

Why defend?

If greenfield sites existed along the complete coastline, the community would not be faced with the dilemma of defending urban properties from the sea. It is arguable that there may not be a need to protect all the fertile coastal strip from sea flooding. However, that is not the case and coastal engineers are required to design effective and acceptable defences to protect both property and land.

Many ancient coastal communities were formed to provide a first line of attack against foreign invaders. They were also inhabited by locals who one way or the other earned their living from the sea. Most of these communities have now developed into large resorts providing holiday accommodation for tourism and recreation - a major industry.

In addition, as a result of uncontrolled development just before and after the Second World War, many unauthorised dwellings appeared in the coastal strips, many on the shingle ridge itself. In more recent times these relatively poor dwellings have been up-rated into very expensive homes which are far from easy to protect against the sea.

While there is still a major need to protect high class agricultural land in the coastal strip against inundation from the sea, there is also a requirement to protect and enhance nature conservation. There is a developing relationship between the NRA and English Nature (EN) to ensure that environmentally sensitive areas such as Sites of Special Scientific Interest (SSSIs) remain. Nature does not stand still and in some instances can be given a helping hand by sensitively designed works. With pressure on the coastline, it is likely that there will be more of these partnership projects.

Project promotion

The history and span of time in the formation of the coastline is very short in comparison with geological time. Man's involvement is even shorter and therefore the promoter of any project that will have an impact on the natural coastal processes must proceed with great care and understanding. There is a need to appreciate the complex interplay between topography, bathymetry, wave climate, prevailing sea movement, surges and so on before any proposals are considered. One of the great concerns of coastal engineers is the lack of specific data that is available for an identifiable area. Money spent on monitoring the coastline is well spent.

With great foresight the predecessor authorities to the NRA commenced aerial survey along parts of Southern Region's coastline in 1973.

This was a joint venture between the then Sussex River Authority, the Crown Estate Commissioners and most of the Maritime Local Authorities. The main objective of the survey was to establish profiles of the foreshore under stable conditions and to provide data which would allow volumes of beach material to be calculated and to establish any major changes. This was achieved by analysing the aerial photographs by photogrammetic means and linking the results to Ordnance Datum. The consortium sponsoring this work contributed towards the costs in return for the data provided.

The monitoring project was gradually expanded so that by 1983 virtually all of the Southern Region coastline was covered. The existence of this type of data is almost unique and provides an excellent starting point for analysing coastal processes. The information is stored as a combination of paper and computer disk systems and is now in need of being incorporated onto a Geographical Information System (GIS). The NRA has entered into a Pilot Study with Babtie Dobbie Limited, to develop such a system. A full Coastal Management System, which it is planned will involve Maritime Local Authorities and others, is planned to be developed starting in 1993. This is an ambitious programme and will require full co-operation of all participating authorities and hopefully will be eligible for grant aid from MAFF.

More and more the strategic approach to sea defence projects is being appreciated. Works must be considered within coastal cells as opposed to individual projects. With the encouragement of MAFF, many coastal groups have been formed and there is now formal liaison so that mutual problems with the coastline can be solved. Joint schemes which encompass both sea defence and coast protection stretches are now a reality.

Having collected the basic data for a full scheme appraisal it is important that all options which offer a solution are studied. The do nothing option should not be ignored. The promoter of a scheme must ensure that the various options are accurately costed and that all the benefits and disbenefits are stated. The worthwhileness of the scheme in benefit/cost terms must be established and the impact of the proposals on the environment assessed. In many cases a full environmental impact assessment will be required and works modified to gain acceptance. To comply with the Treasury Guidelines, it may be necessary to undertake an economic appraisal of the various options. In the past, many coastal projects were undertaken solely by engineers. Currently it is essential to include both engineers and environmental experts in the team. In the future the project team will have to be fully multi-disciplinary to include engineers, environmental experts, economists and landscape architects.

During the scheme preparation it is very necessary to keep the 'approving' authorities involved. If the project is subject to grant aid from MAFF, then the Regional Engineer should have a place in the project team.

Scheme development

In the past, managing the coastline has been undertaken by a multitude of coastal authorities. It is only in recent years that MAFF has assumed a co-ordinating role between sea defence and coast protection projects, so that consistent standards of protection are achieved. Although there must be a consistency of approach in analysing the problem, each preferred solution will be identified on its merits.

Coastal engineering still relies a great deal on judgement, and while modelling techniques today can help, there is usually more than one way to succeed. In the past beach replenishment, groyning and re-cycling have been successfully undertaken within Southern Region. More recent studies have revealed that on some stretches of coastline, the use of rock in the form of revetments, bastions and rock islands may reduce the future maintenance costs of defences. Although it is early days, further studies will confirm the construction elements to be included.

Case histories

Many successful schemes have, over the years, been promoted in Southern Region, some of which are outlined below with comments on their performance.

Seaford sea defences - shingle replenishment

Although there were a number of elements to this scheme, the most significant part was the replenishment and re-profiling of the Seaford beach over a 2.5 km stretch. Some 1.5 million cubic metres were placed on the beach, imported by sea. The material was won by suction dredger from the Owers Bank with the agreement of the Crown Estate Commissioners, and then pumped ashore. Work was completed in just over six months and involved 24 hour working. The scheme was substantially completed just before the storm of October 1987. Although the storm occurred at a period of neap tides, little damage was observed.

In order to maintain the replenished beach profile, it was assessed that some 20 000 to 25 000 cubic metres of recycling of material would be necessary. While this has varied from year to year, currently some 50 000 cubic metres of material is re-cycled and re-profiled. Regular monitoring of the beach has been undertaken and recent indications are that the beach is stable in relation to quantity and if anything shows a slight accretion.

Pett sea defences - shingle replenishment

This scheme, undertaken in the early 1950s, was necessary to reinforce the failing concrete sea wall between Fairlight and Winchelsea. The littoral drift was from west to east resulting from the prevailing south-westerly gales and shingle accumulated at Nook Point at the entrance to Rye Harbour. The solution adopted was to recycle by lorry some 150 000 cubic metres of material in a westerly direction, to reinforce the old seawall and to raise the beach crest level to give approximately a 250 year protection. As part of the proposal it was necessary to undertake annual re-cycling of 20 000 cubic metres of shingle. The scheme has proved successful but in recent years the amount of material required to be re-cycled has increased. A reappraisal of this length of coastline is under way.

Dymchurch sea wall - hard sea wall

This defence, which takes in Dymchurch, St Mary's Bay and Littlestone, is mainly a hard defence and has been developed and reinforced over many centuries. Over the years, works have been undertaken to combat falling foreshore levels and experiments with various groyne fields have been tried without success and many of the groynes have now been allowed to fall into disrepair.

In the 1960s, after model test at HR Wallingford, major reconstruction was undertaken with the toe of the defences being moved seaward by some 6 metres to combat further lowering of the foreshore. After storm damage in 1990, further works were necessary above the normal maintenance to replace parts of the coast. In 1990 consultants were appointed to undertake a strategic study of the defences to promote a long-term investment programme. Some parts of the defence were found to be in urgent need of repair. Works are ongoing to reconstruct the concrete crest wall at Grand Redoubt, with further works planned for the medium term.

Camber Sands - sand dunes

Over the years Camber Sands have been very popular with holiday-makers and the continuing pedestrian traffic was damaging the sand dunes which offered an effective sea defence. In 1964 a jointly funded programme of stabilisation works was undertaken which included sand fencing and the planting of marram grass. This has proved to be very effective and the dunes have now been re-established. While the need for some of the sand fencing has diminished, there is still a real need for clearly defined access ways. Prickly buckthorn is a very useful deterrent to sunbathers.

Pennington sea wall - hard defence

The need to reinforce the sea defences that protect Pennington Marshes between Keyhaven and Lymington was graphically demonstrated by the flooding that occurred in 1989, when Lymington was severely flooded. The proposal to reconstruct the Pennington sea wall had been taken prior to the flooding and phase I of the works was already under way. Phase II was planned to be undertaken in spring 1991, but was delayed by the inability of the NRA to obtain a licence from English Nature (EN) to undertake works within an SSSI. The works, as planned, required the relocation of the habitats of two listed species - shrimps and sea anemones - under the Wildlife and

Countryside Act 1981. Although EN were appreciative of the NRA's needs, legally they could not grant a licence. After much delay and modification to the scheme, the first licence to be granted by MAFF under Section 16 of the Act was obtained. Phase II started in November 1991 and will be substantially completed by the end of 1992.

Conclusions

To provide an effective defence the beaches that surround our coastline will continue to need engineering. The supply of natural shingle for replenishment appears to be finite and will need to be replaced with marine dredged material which in itself appears to be finite. There will be increasing competition from construction aggregate firms for this scarce resource and there is a need to establish whether or not non-aggregate materials will be suitable for beach replenishment. Research into this is being conducted by the NRA with assistance from the Crown Estates Commissioners.

Further development of Coastal Zone Planning is essential especially with the advent of sea level rise. Scheme appraisal must consider all options including the do nothing solution. Although not necessarily applicable at present, serious consideration will have to be given to the retreat option which could provide reinforcement to the fast diminishing saltings, the habitat of many wading birds.

Acknowledgements

The author wishes to thank staff of Southern Region NRA for their help in providing the information for this Paper. The author's views expressed in the paper are not necessarily those of the NRA.

References

1. Stokes C. Romney Marsh Levels. Memorandum (1930).
2. Southern Water. Internal Document. Beach Monitoring Summary (1989).
3. Holmes A.E. Seaford Sea Defence Scheme. IWEM (1988).
4. Robinson G.W. Beach Recharge. ICE Maritime Engineering Group (1989).
5. Shave K. Beach Management. ICE (1989).

21. Engineering with conservation issues in mind

K. A. POWELL, BSc, PhD, Manager, Coastal Management Section, HR Wallingford

It has often been considered that engineering objectives and conservation requirements are largely in conflict. However, by seeking a full understanding of each other's ambitions, coastal defence systems can be engineered satisfying more than one requirement. A clear example of this approach is seen in a recent study of types of coastal defence method that are appropriate for various locations experiencing cliff erosion, where continual exposure of the geological sequence in the cliff is of great signficance to the conservationist. By minimising the rate of erosion compatible with exposure of the cliff face, a defence system can often be devised for a location which will satisfy both the conservation and engineering requirements. Such an approach is outlined in this paper.

Introduction

Many classic geological localities in Britain take the form of coastal cliffs. These, by virtue of their international, national or regional significance, have been recognised as SSSIs or RIGS. Their value as research and teaching sites relies on the exposures being kept fresh by continued marine erosion. As a consequence many of the sites are inherently unstable, with rates of erosion of up to 2-3 m per annum (ref. 1).

Where such sites occur close to areas of habitation or recreation, there is often intense local pressure to find engineered solutions to the problems of coastal recession. Traditionally, these solutions take the form of concrete sea walls or rock revetments, often coupled with cliff grading or drainage. In the extreme, such schemes may lead to the total obliteration of the geological interest of the site. Even relatively small-scale works can inhibit the action of the sea, and thus prevent marine erosion, leaving the cliffs to become stabilised and overgrown within a few years. Once stabilised the resultant slope is of little value as a teaching and research site, as the underlying strata are no longer readily accessible. Furthermore, there is usually a marked reluctance on the part of the landowner to allow any subsequent geological sampling, as this would necessarily involve disturbing the equilibrium of the stabilised slope.

ENGINEERING SYMPATHETIC SOLUTIONS

In an attempt to resolve the conflicts between current engineering practice in coast protection and conservation, the Nature Conservatory Council (now English Nature) commissioned HR Wallingford to review alternative coast protection techniques and novel applications of traditional structures. The objective of the study was to identify engineering options which would provide partial stabilisation of the cliffs, sufficient to ease the concerns of landowners, whilst still maintaining the geological interest of the site.

This paper sets out the main proposals from the HR study (ref. 2) with regard to the development of a methodology for the selection of appropriate coast protection works for geological SSSIs. The methodology has four main elements

(a) site appraisal and classification
(b) identification of suitable protection techniques
(c) application of protection techniques to different site categories
(d) selection of appropriate protection strategies.

The application of the methodology, as outlined in this document, will allow suitable compromise coast protection options for geological SSSIs to be identified, though not designed. Inevitably, there will be some sites for which compromise solutions cannot be found: in these situations it is hoped that this study will at least serve to promote dialogue between the various parties involved.

Coastal management and SSSIs

Historically, coastal management in the UK has been on the basis of administrative boundaries. However, these boundaries do not usually coincide with the natural coastal cells - that is sections of coastline (usually a bay) self-contained to the extent that whatever happens in one cell does not necessarily affect the next, or neighbouring cells. Thus a district authority might have several cells within its boundary or, alternatively, several authorities might be involved in one cell.

Many of the problems that arise on the coastline are a direct result of a piecemeal, unco-ordinated, approach to coast protection, in a manner that ignores the natural coastal process boundaries. They can be exacerbated when there are a large number of bodies, such as commercial concerns, the Ministry of Defence, The National Trust, British Rail, private land owners etc, involved in the administration of parts of the coastline: Often these various bodies have their own particular interests and concerns which may not be in accordance with those of their neighbours.

It is clear that the way forward lies in managing the coastline as a series of discrete, self-contained cells with the boundaries fixed by natural rather than administrative processes. The first steps down this path have already been taken with identification and documentation of the various cells, and sub-cells, around the UK coast (ref. 3).

Administering the coastline on the basis of coastal cells would ensure not only the most cost-effective employment of coastal defences but would also allow the natural defences to be improved, and brought more fully into our coast protection armoury. Assessment of the impact of the proposed scheme on beach levels throughout the local area is clearly important. If, for instance, the works would act to retain or impede the transport of beach material, then this could have a detrimental effect elsewhere on the coast. A similar situation may arise if the cliffs to be protected are the source of substantial quantities of beach-forming material. In these instances the do nothing option might be the most appropriate strategy. In such circumstances it might be beneficial to view important geological sites as beach-forming material as they erode and retreat.

Coastal geological SSSIs
Coastal cliffs

The largest group of coastal geological SSSIs are those located in coastal cliffs. Unfortunately, by their very nature, many of these sites occur on eroding coastlines where there is often local pressure to prevent continued recession. Many of the protection schemes adopted are, however, detrimental to the geological interest of the site, either because they obscure the rock sequences directly or, more importantly, prevent the slow erosion necessary to maintain exposures in softer deposits.

Intertidal sites

Many of the geological coastal SSSIs have part or all of their interest located in the intertidal zone. The threats likely to be experienced by these sites are, however, distinctly different from those affecting coastal cliff SSSIs. Generally erosion in the intertidal area is of little concern to the local authorities and it is therefore rare for these sites to require protective works. The problems are instead associated with adjacent works or developments which either directly obscure the intertidal interest or else alter the local littoral regime in a way that is itself detrimental to the SSSI. Examples of possible threats to intertidal SSSIs include

(*a*) the development of marina or harbour facilities
(*b*) reclamation works
(*c*) the construction of tidal barrages
(*d*) replenishment of adjacent beaches
(*e*) the undertaking of local works that may themselves promote beach secretion or siltation within the SSSI.

ENGINEERING SYMPATHETIC SOLUTIONS

In all of these cases the overriding need is to ensure that the detrimental effects of any works are mitigated as far as possible and that access is maintained to the geological exposures.

Estuarine sites

SSSIs located in very sheltered estuaries or someway up river are not usually subject to serious wave activity. Instead most of the marine erosion arises through the actions of tidal or river currents and ship wash. At these sites therefore a different set of coast protection works will be required, with the emphasis being largely on diverting currents away from the base of the cliffs.

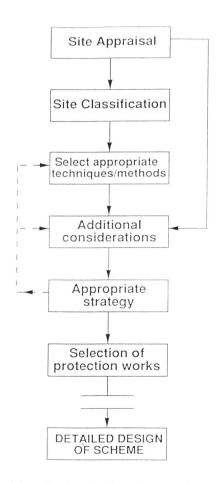

Fig. 1. Proposed methodology for the selection of appropriate coast protection works

The selection of appropriate coast protection techniques for geological SSSIs

Methodology

The methodology proposed to enable the selection of appropriate coast protection works in geological SSSIs is outlined in Fig. 1. Essentially there are three main stages in the process, following the initial site appraisal

(a) the classification of the site

(b) selection of possible protection techniques from which appropriate site specific solutions can be subsequently derived

(c) selection of a strategy within which the preferred scheme can be applied.

These items are considered in more detail in the following sections.

Site appraisal and classification

Although each SSSI is unique either in terms of geological interest, or location, or environmental conditions, it is possible to develop a general classification for the sites based upon similarity of a number of features. The advantage of such a classification is that it allows the development of methodology for the selection of appropriate coast protection works to be couched in more general terms than would otherwise be possible.

The site classification employed in this document is shown in Fig. 2, and is based on an assessment of

(a) the nature of the geological interest (palaeontological, stratigraphic, coastal geomorphological or mass movement)

(b) the location of the interest (cliff or intertidal foreshore)

(c) the causes of erosion (marine, groundwater or a combination of both).

In the development of this classification a number of assumptions have been made.

(a) Sites which contain only a stratigraphic interest, and which are subjected only to marine erosion, can be treated separately since they generally require a lower level of marine action to maintain the geological interest (i.e. sufficient to clean faces, remove talus etc) than palaeontological or mass movement sites.

(b) Intertidal sites are subject only to marine erosion.

(c) The erosion processes in mass movement sites are due to a combination of both ground water seepage and marine action.

In addition it should be noted that the sites have been classified in accordance with their current condition, regardless of whether or not this is typical of their long term undisturbed state.

ENGINEERING SYMPATHETIC SOLUTIONS

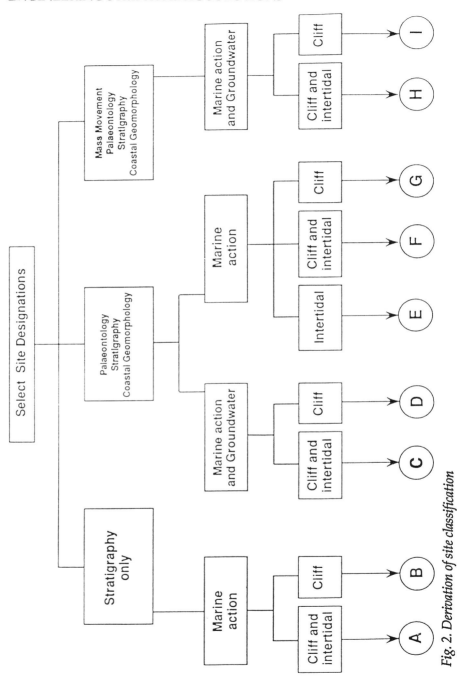

Fig. 2. *Derivation of site classification*

The resulting classification yields nine categories of site which may be summarised as

(a) combined cliff and intertidal sites containing only a stratigraphic interest and subject only to marine erosion
(b) as (a) but with the interest confined to the cliffs
(c) combined cliff and intertidal sites designated for either stratigraphic, palaeontological or coastal geomorphological interests, or any combination thereof; erosion by both marine action and groundwater seepage
(d) as (c) but with the interest confined to the cliffs only
(e) as (c) but intertidal sites only and with erosion due solely to marine action
(f) as (c) but with erosion due solely to marine action
(g) as (c) but cliff sites only and with erosion due solely to marine action
(h) cliff and intertidal sites containing a mass movement interest; erosion due to combined effects of marine action and groundwater flows
(i) as (h) but cliff sites only.

Since many SSSIs cover large stretches of coast, they may also cover a range of geological types and interests. It is likely therefore that some of the larger SSSIs will include areas with different classifications. If this is the case it is recommended that when problems arise at a particular location within an SSSI, that location is re-classified according to its distinct geology and conditions.

Protection techniques

Coastal cliff sites. There are a number of fundamental techniques which may be adopted, either in combination or singly, to provide protection to eroding coastal cliffs. Many of these may, however, be detrimental to any geological interest and therefore unsuitable for use in designated SSSIs. Where the techniques require the deployment of specific structures (breakwaters, seawalls etc) there may be a number of different structural types and materials that can be used.

The various techniques and methods that may be employed in coast protection works in SSSIs are listed in Table 1. The Table is broken down into a number of protection techniques (i.e. land drainage, wave attenuation etc) and, where appropriate, methods associated with those techniques. The methods themselves may also fall into a number of different categories; thus for direct protection by static structures, for example, the appropriate methods may be classified as either seawalls, revetments, timber palisades, or saltings.

Although Table 1 does not provide an extensive list of structure types, it does cover most of the general categories of structure, distinguishing, for example, between vertical and sloping concrete sea walls. A few of the

ENGINEERING SYMPATHETIC SOLUTIONS

'novel' structures have been omitted from the list because their use was felt to be unproven or inappropriate within the context of this study.

Techniques for intertidal sites. If, in case of intertidal sites, it cannot be guaranteed being obscured then it will be necessary to provide for continued access to the site. This access could take the form of caissons built round areas of particular interest. However, if this is to be a cost effective approach the interest would either have to be concentrated in a number of specific locations within the site, or the site itself would have to be relatively small. An alternative approach for some sites would be to enter into a management agreement with the landowner/developer whereby the site would be peri-

Table 1. Appropriate protection techniques and methods

Land drainage	Surface drains
Deep drainage systems	Horizontal drainage systems
	Drainage galleries, adits or tunnels
	Walls and vertical drains
Wave attenuation	
Offshore breakwaters/sills	Rock
	Concrete armour units
	Sunken vessels
Direct protection - static structures	
Sea walls	Vertical concrete
	Gabion baskets
	Sloping concrete
	Vertical sheet pile
Revetments	Rock
	Concrete block
	Asphaltic
	Scrap tyres
	Timber palisades
	Saltings
Direct protection - dynamic structures	
Beaches	Sand beach (with groynes)
	Shingle beach (with groynes)
	Rock beach
Strongpoints	Rock
	Concrete armour units
Cliff strengthening	Infill concrete walls
	Infill blockwork walls
	Rockbolts and mesh
Purchase or compensation	
Re-creation of interest inland	
Do nothing	

odically cleared of deposited sediments, or enforced water levels periodically dropped, thus allowing access to the designated exposures.

Techniques for estuarine sites. SSSIs located in very sheltered estuaries or someway up river are not usually subject to serious wave activity. Instead most of the marine erosion arises through the actions of tidal or river currents and ship wash.

Designing protective works to counter the effects of ship wash is relatively straightforward, and most of the systems previously listed for wave protection may be considered provided they do not represent a hazard to navigation. In the case of current induced erosion there are two possible approaches

(a) provide direct protection to the cliff toe
(b) divert the currents away from the cliff.

The options for the first of these approaches are much the same as those outlined for coastal cliffs. The second category of works, however, would require a number of shore normal structures (training walls or groynes) to be built. The length and spacing of these would need to be optimised in further design studies but would basically depend on the size of the site, current velocities etc. At some sites shore normal structures may need to be combined with direct protection works in order to reduce cliff recession to an acceptable level, and to counter the development of eddies which may themselves result in additional erosion of the cliff toe.

Application of protection methods to different categories of site

The application of the different coast protection techniques to the different categories of site is shown in Table 2. The selection of suitable protection methods needs to be site specific, and in particular to distinguish between

(a) sites where the cliffs may or may not be prone to rock falls or slippages, which may cause damage to protection works or cause a safety problem during construction
(b) sites where all or part of the interest is located at the base of the cliffs, and therefore may be obscured by inappropriate protection works, and those where the interest is located elsewhere in the cliff face.

Additionally it is necessary to identify those methods which may be appropriate to a low technology approach using unskilled labour and employing locally available tools and materials.

Table 2 presents the main output of the study, but it can only be used by the design engineer as a starting point. Taking the selection of a coast protection strategy to the more advanced stage of detailed design involves the consideration of many site-specific factors, such as hydraulic conditions and geographic location. It should be noted, however, that engineering

ENGINEERING SYMPATHETIC SOLUTIONS

Table 2. Selection of coast protection options

APPROPRIATE PROTECTION METHODS

TECHNIQUES	CATEGORY A AND F SITES				CATEGORY B AND G SITES				CATEGORY C SITES	CATEGORY D SITES			CATEGORY E SITES	CATEGORY H SITES	CATEGORY I SITES	STATUS OF WORKS
	Cliffs prone to falls or slips?		All or part of interest located at base of cliff?		Cliffs prone to falls or slips		All or part of interest located at base of cliff?			All or part of interest located at base of cliff?						
	Yes	No	Yes	No	Yes	No	Yes	No		Yes	No					
1 Land drainage:																
Surface drains									*	*	*					Low technology
Deep drainage systems																
Horizontal systems									*	*	*					
Galleries, adits or tunnels									*	*	*					
Wells and vertical drains									*	*	*					
Trench drains									*	*	*					
2 Wave attenuation:																
Offshore breakwater/sills																
Rock	*(1)	*(1)	*(1)	*(1)	*	*	*	*	*(1)	*	*		*(1)	*		Low technology
Concrete armour units	*(1)	*(1)	*(1)	*(1)	*	*	*	*	*(1)	*	*		*(1)	*		
Sunken vessels	*(1)	*(1)	*(1)	*(1)	*	*	*	*	*(1)	*	*		*(1)	*		
3 Direct protection - static structures:																
Sea walls																
Vertical concrete		*(1)	*(1,2)	*(1)		*	*(2)	*								
Gabion baskets		*(1)	*(1,2)	*(1)		*	*(2)	*								Low technology
Sloping concrete						*	*(2)	*								
Vertical sheet pile		*(1)	*(1,2)	*(1)		*	*(2)	*								
Revetments																
Rock					*		*									Low technology
Concrete block (4)					*		*									
Asphaltic					*		*									
Scrap tyres (3)					*		*									Low technology
Timber palisades	*		*(1,2)	*(1)		*	*(2)	*								
Saltings (4)					*	*		*					*		*	Low technology
4 Direct protection - dynamic structures																
Sand beach (groynes)						*		*					*		*	
Shingle beach (groynes)						*		*					*		*	
Rock beach						*		*					*		*	Low technology
5 Strongpoints:																
Rock		*	*(6)	*		*	*(6)	*	*(6)	*(6)	*					Low technology
Concrete armour units		*	*(6)	*		*	*(6)	*	*(6)	*(6)	*					
6 Cliff strengthening:																
Infill concrete walls	*(7)	*		*	*(7)	*		*								
Infill blockwork walls	*(7)	*		*	*(7)	*		*								
Rockbolts and mesh	*(7)	*	*		*(7)	*	*									
7 Purchase or compensation	*	*	*	*	*	*	*	*	*	*	*		*	*	*	N/A
8 Re-creation of interest inland	*	*	*	*	*	*	*	*	*(5)	*	*					Low technology
9 Do nothing	*	*	*	*	*	*	*	*	*	*	*		*	*	*	N/A

Notes

1 Careful siting required to avoid obscuring any intertidal interest
2 Remote from base of cliff to allow access to lower strata
3 Very sheltered locations only
4 Low or moderate wave climates only
5 If suitable natural sites exist
6 Will obscure short sections of interest
7 Safety consideration during construction

works will normally result in some loss of interest, so there can be no guarantee that protection works will be compatible with effective site conservation in specific instances.

Additional considerations

From Table 2 we have a range of coast protection techniques and methods that may be suitable for a particular category of SSSI. However, just as the sites grouped together under a particular classification will differ in their location, geography, requirements etc, so too will the protection techniques and methods that are most suited to them. Thus in addition to the considerations which relate specifically to the nature of the site and the natural forces acting on it, there are a number of other factors to be taken into account in deciding on the most appropriate coast protection strategy. These include

(a) the general nature of the coastal cliffs
(b) the proximity of existing or proposed development
(c) additional site designations
(d) cost-benefit requirements
(e) construction methods and access
(f) integration with existing works
(g) implementation within a coastal management stategy.

Selection of appropriate protection strategy

The final stage in the development of a cost-effective protection methodology for geological SSSIs is the selection of an appropriate strategy within which the engineering solutions outlined in Table 2 may be implemented. The options may be summarised as follows.

S1. Permanent partial protection to entire SSSI
S2. Permanent partial protection to selected lengths of SSSI - controlled erosion
S3. Permanent full protection to selected lengths of SSSI - controlled erosion
S4. Permanent full protection to entire SSSI including provision of access windows and/or inspection chambers

Of these, strategy S1 requires that the entire site is afforded a level of protection sufficient to reduce recession rates to a level acceptable to the land owner whilst still ensuring continuity of the geological interest. The problem with this approach is one of design since it is extremely difficult to relate cliff recession rates directly to a reduction in wave action, without first undertaking either extensive hydraulic modelling of the scheme, or carrying out prototype measurements in the field. Nevertheless in practice this is probably the approach that will be most often adopted.

Strategy S2 is subject to similar problems but to a lesser scale, since the partial protection is only provided to limited areas of the SSSI. The remaining areas are then allowed to continue to erode and to gradually form a number of small 'stable' embayments, with less land loss than would have occurred had the whole site continued to erode at its previous rate. One of the advantages of this approach is the flexibility in the siting of the strongpoints, allowing them to be located in areas where the erosion is most rapid or where its consequences are most severe. Strategy S3 is similar to S2 but offers full protection at the strongpoints. Overall recession rates are therefore likely to be less than those for S2.

Strategy S4 would be appropriate for stratigraphic sites where it might be acceptable either to install inspection chambers within sea walls, allowing access to otherwise obscured strata, or to maintain access to selected lengths of the cliff face by periodically cleaning away talus. Generally, this strategy will only be realistic for sites where the interest is concentrated in one or two areas.

Table 3 identifies the strategies appropriate to the various site classifications and protection techniques. It should be noted that strategies are only appropriate to engineered coast protection schemes and that careful consideration should be given to the operational requirements of each strategy.

Summary

This paper provides an overview of the coast protection techniques currently available and their applicability as the least damaging options for particular categories of geological site. The methodology outlined is not a substitute for a detailed site-specific study nor does it guarantee that the coast protection techniques described will be consistent with effective site conservation in every situation. It will, however, be a useful starting point

Table 3. Selection of coast protection strategies

Site Classification	A	B	C	D	E	F	G	H	I
Strategy :	S1	S1	S1	S1	–	S1	S1	S1	S1
	S2	S2	S2	S2	–	S2	S2	S2	S2
	S3	S3	S3	S3	–	S3	S3	–	–
	S4	S4	–	–	S4	–	–	–	–

Technique	T1	T2	T3	T4	T5	T6	T7	T8	T9
Strategy :	S1	S1	S1	S1	–	–	n/a	n/a	n/a
	S2	S2	–	–	S2	–	n/a	n/a	n/a
	S3	S3	–	–	S3	–	n/a	n/a	n/a
	–	S4	S4	S4	–	S4	n/a	n/a	n/a

for discussions between coast protection authorities and conservationists as and when conflicts arise over specific proposals.

References
1. McKirdy A.P. Protective works and geological conservation. In Planning and Engineering Geology by Culshaw, Bell, Cripps and O'Hara, 81-85. Geological Society (Engineering Geology Special Publication No. 4).
2. Hydraulics Research Ltd. A guide to the selection of appropriate coast protection works for geological sites of special scientific interest. Report No. EX 2111, February 1991.
3. Hydraulics Research Ltd. A macro-review of the coastline of England and Wales. Vols 1 to 8.

Discussion

There was general acceptance of the merits of beach nourishment as a soft solution to coastal defence. The use of beach nourishment and the possibility of roll-back solutions were the subject of research, including the use of dredged seabed materials which did not compete with the offshore aggregates industry for reasons of location, grading or contamination.

Some doubt was expressed as to the use of inshore materials on the basis of the overall balance of benefit. Moving inshore might merely exchange new problems for old and introduce conservation issues. Finer materials stood at a flatter gradient and therefore required a greater volume. Dredging industry costs for such material would have to be very competitive to compete with more traditional gradings.

Dredged spoil had been used in South Africa for beaches as a tourist asset. Further investigation into the similar use of dredged spoil in the UK was suggested in order to assess the effect of contamination. It was already used at Bournemouth and was in any case not always welcomed by conservationists for dumping offshore.

The quality, shape and grading of seabed material was critical and required thorough investigation which was very costly. The possibility of a coastal group obtaining a licence for its own area was discussed but prospecting and licensing was a slow process.

Reference was also made to the use of sediment bypassing at harbour entrances, to the safety problems arising from the use of rock revetments on beaches and to the value of some hard defences as a back-stop to beach nourishment schemes under severe draw-down conditions.

22. The Anglian Management Study

M. W. CHILD, BSc, MICE, MIWEM, Engineering Manager, National Rivers Authority, Anglian Region

The objective of coastal zone management is to understand and resolve conflicts and move forward to integrated solutions. Such an integrated approach particularly in the Anglian Region, where 70% of the coastline is defended and an area of land equivalent to the size of Essex is at or below flood risk level, requires an understanding of coastal processes and sound strategic plans. This paper describes the inception of the Anglian Managment Study in 1987, the costs, the benefits and its application to the wider aspects of coastal zone management.

Inception and development
It was against a background of

(a) a real risk of flooding
(b) a justified need for major flood defence investment
(c) a geomorphologically and geologically complex and diverse coastline
(d) the possibility of sea level rise from global warming

that the Sea Defence Management Study (SDMS) was conceived in 1987.

After completion of a preliminary study the SDMS was commenced in 1988. By the spring of 1991, after an investment of £1.65 million and three years of effort, the SDMS was completed. This, however, is only the beginning. The study has introduced new concepts and a new approach to the planning and provision of sea defences. Although this approach is in its infancy, the benefits derived are already significant. With continued development the approach will be of increasing benefit for effective flood protection of the Anglian Region and a positive contribution to the wider field of coastal zone management.

The Anglian Region
Much of the Region is flat, lowlying and below maximum recorded sea level. The Region covers one of the most vulnerable and variable coastlines in Britain stretching from the Humber in the north to the Thames in the south.

With over one fifth of the Region below flood risk level a major investment programme is planned. To ensure the continuing protection of three quarters of a million people and the billions of pounds of investment in infrastructure

CASE STUDIES

and land, a ten year programme amounting to a total expenditure of £340 million on coastal and tidal defences is being implemented.

The Region is protected from tidal flooding by about 1500 km of defences. The wide range of coastal geomorphology, the underlying geology and the exposure to wind and waves require a variety of defence solutions for this sensitive coastline. Coastal towns and land are protected by natural systems such as sand dunes and beaches, and engineered defence structures such as groynes, embankments and sea walls. Estuary towns and land are protected by saltings, embankments, flood walls and surge barriers.

There is evidence that flood defence was carried out as early as the Roman times, but the major reinforcement and extension of the defences took place after the disastrous 1953 east coast flood in which over 200 people drowned. Limited refurbishment also took place in those places affected by the 1978 flood, which caused major damage to Wisbech, King's Lynn and parts of the Norfolk coast; fortunately this time with no loss of life. On both occasions the North Sea surge was the major factor contributing to the damage.

North Sea surges

Surges occur frequently in the North Sea. They are created by low atmospheric pressure and the funnelling effect of the North Sea coastline. Surges of 1 m occur four or five times a year but often not coincidental with spring tides, onshore winds or high tide. During a surge the sea level rises dramatically and in the case of 1953 the sea rose over 2 m higher than predicted and in 1978 the still water levels recorded were higher that the 1953 level.

The need to understand processes

Protection against surges dictates the level, the location and often the type of defence. However, it is the ongoing day in day out rise and fall of the tide, the tidal currents and the action of wind and waves on the coast that can undermine defences, cause accretion and erosion and change the shape of the foreshore. Effective flood defence in the Region therefore requires knowledge of processes and changes along the coastline as a whole on an integrated basis irrespective of responsibility. Such an approach has been adopted in the SDMS. It is an approach that is not only very relevant today but even more relevant tomorrow with the possibility of global warming and sea level rise.

Study area

It was felt necessary to extend the northern boundary of the study to incorporate the Holderness coast which includes a potentially important sediment source for the East coast. In contrast the Kent coast, on the south side of the Thames Estuary, was not considered to interact significantly with the regime along the Essex coast and was not therefore included. The extent of the study area is shown in Fig. 1.

The preliminary study (1987 to 1988)

The preliminary study marked a major initiative to move away from the piecemeal approach of the past towards an integrated look at the whole coastline. An early task was to examine the feasibility of this approach, and to identify topics for investigation and to consider the cost implications. In late 1987 Halcrow was appointed for the preliminary study

(a) to examine and collate references and data sources for the coast
(b) to identify dominant coastal processes and responses
(c) to develop a short term management strategy

This stage of the study was completed in 1988 at a cost of £0.25 million. The output of the study included (ref. 1)

(a) a coastal atlas showing coastal data and information pictorially
(b) a database of references and sources of information

Fig. 1. The study area and flood risk area

CASE STUDIES

(c) a Geographical Information System (GIS) of the main variables
(d) a study report
(e) a short term management strategy.

Benefits from the preliminary study were (ref. 2)

(a) a better understanding of the coastal processes (e.g. offshore bank formation)
(b) a better understanding of responses (e.g. steepening occurring over 70% of the coastline)
(c) a database of references and facts invaluable for strategic management and scheme development.

The completion of SDMS (1988 to 1991)
During 1988 Halcrow was appointed with the following tasks

(a) filling in data gaps of the preliminary study
(b) continuing the further development and understanding of the coastal processes and mechanisms
(c) refining and developing the Geographical Information System into a fully operational tool
(d) defining a monitoring programme
(e) defining a management strategy.

The overall objective was to develop a management strategy on a sound basis for investment plans for flood defence.

Approach
As already noted the need to obtain some basic understanding of coastal mechanisms along the Anglian coast was seen as the foundation on which to build the management strategy. In order to achieve this it was considered necessary to

(a) define the management framework
(b) undertake studies to provide an understanding of the coastal processes involved
(c) initiate a programme of field work required to further this understanding
(d) establish a system by which information can be extracted, manipulated and updated by those implementing the management strategy.

The approach adopted was to construct a picture of the dominant processes from the wide range of information that was already available. This philosophy was adopted because it was felt that over such a large and diverse area, any attempt to apply numerical models to examine processes and coastal development would inevitably be constrained by limited knowledge of the governing processes and mechanisms. The prime objective was therefore to structure the data in such a way that it could be rapidly manipulated in order to develop understanding.

A thorough analysis of existing sources of information was then made, both to gain insights and to focus susequent field work and numerical model studies. In this way both the information base and the interpretive capability were progressively improved.

Study components

In order to achieve this the project was divided up into a number of task areas as follows.

(a) *Data collection.* Initially this focused on extracting information from existing archives. Once the most significant gaps had been identified, a major field survey programme completed the required data coverage.

(b) *Supporting studies.* These ranged from studies to exploit existing data sources (e.g. wave hindcast, analysis of extreme sea levels), through the development of models to investigate specific phenomena (e.g. a beach response model to investigate beach steepening), to impact studies (e.g. sea level rise and climatic change) and reviews (e.g. literature, changes in the North Sea basin and Essex Saltings Programme).

(c) *Monitoring programme.* A number of studies investigated various aspects of monitoring to provide the basis for defining a comprehensive regional monitoring programme. In addition a software package was developed to support the collection and subsequent analysis of the data.

(d) *Management system.* This entailed the development of a geographic information system (GIS) to meet the needs of the approach as outlined above.

(e) *Management strategy.* Based on an analysis of the governing coastal processes in conjunction with the Authority's management objectives, the basis for response options can be evaluated. More importantly, the strategy forms one part of the management framework which comprises data collection, information management, and strategic planning.

Whilst the database includes sea defence, this does not include historical developments, nor any detailed investigations into structural integrity. Furthermore the development of the management strategy is based on a technical evaluation of the available data and does not include any benefit-cost considerations. Both of these aspects were outside the scope of this project.

The various components and findings of the study were documented in a series of 25 reports (refs 3 and 4).

CASE STUDIES

Costs and benefits

Introduction

The decision in 1987 to embark on an integrated coastal study was farsighted. The SDMS has now come to a very successful fruition with many benefits. The timing has coincided with unprecedented interest in the coastal environment, the issue of global warming, climate change, sea level rise and coastal zone management. It provides a sound basis for discussion and informing the public and all those in the coastal zone on these issues.

The cost

18. The total cost of the SDMS was £1.65 million. The breakdown of the total cost is approximately

	£million
Preliminary Study (1987-88)	0.25
Field Measurements	
Bathymetric survey	0.40
Nearshore geological survey	0.15
Estuary sediment trends	0.18
Nearshore currents	0.16
Sediment modelling	0.10
Offshore banks	0.02
Impact of climate change	0.02
Impact of sea level rise	0.02
Estuary studies	0.03
Monitoring	
Beach survey methods }	
Satellite data }	0.17
Monitoring programme }	
GIS and management strategy	0.24
Total	1.65

(This cost excludes the purchase of GIS hardware at approx £40,000.)

Benefits of study

The prime purpose of the SDMS is to provide a sound basis for investment plans. However, to achieve this objective and to ensure that the management strategy is both technically sound and regionally coherent, a diverse and wide range of topics have been investigated. The study is in essence a large research and development project with many of the initiatives being exploratory in nature. As in the case of research and development projects, some parts of the study have been more fruitful whilst others have been less fruitful than expected. Perhaps not fully appreciated at the time of com-

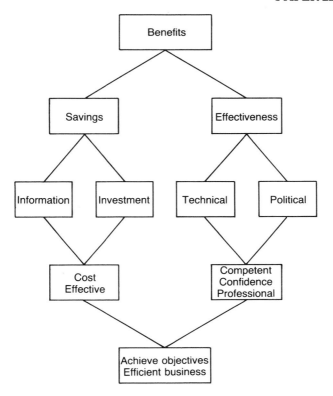

Fig. 2. *Study benefits*

mencement of the study, was the total range of benefits that would ensue from the study.

The benefits of the study can be broadly split as outlined in Fig. 2 (ref. 5).

Savings which will arise from project development are direct cost savings or costs that would have been incurred (tangible benefits). Savings will arise from two areas of project development

(*a*) information: improved quality and availability of data; economy of scale in colleciton and collection

(*b*) investment: investment options more robust and wide ranging.

The effectiveness of flood defences will be improved in two areas

(*a*) technical: improved knowledge and understanding to develop standards, asset plans and levels of service

(*b*) political: to move towards a shoreline management approach; to develop corporate and management plans; to improve image, public

CASE STUDIES

relations and education; to contribute to the wider aspects of coastal zone management.

These benefits produce a cost effective, competent, confident and professional service ensuring the achievement of objectives and an efficient business (see Fig. 2).

Benefits and savings achieved

The ready availability of information on sea defences and the coastline as a whole has already proved invaluable in dealing with the media, environmentalists and conservationists and public interest groups. The NRA is an environmentally aware and open organisation and the study has done much to maintain and improve this image. Most projects require an environmental statement. Data and understanding of processes obtained from the study have already been incorporated in several environmental statements.

Data was provided for the National Sea Defence Survey in 1990 for the whole of the Anglian coastline. This was computer generated from the SDMS database. Without the SDMS a cost of at least £0.3 million would have been incurred on this exercise alone.

During the last year a number of projects have used the SDMS data and information. Projects include the Heacham Hunstanton Beach Recharge (£4.2 million), the Essex Rural Walls (£30 million), Lincolnshire Coast Study (£30 million), Happisburgh/Winterton groynes (£30 million) and numerous smaller projects. It is estimated that without the SDMS at least £0.3 million would have been incurred on data collection and retrieval. The study has ensured easy access to a comprehensive range of good quality data to support these important projects and a total of £0.6 million saving has been achieved.

Benefits and savings

The £0.6 million saving already achieved is only the beginning. A total 10 year investment programme of £340 million on sea and tidal defences will continue to demand extensive and quality data from the SDMS. A simple 10 year savings chart has been produced (Fig. 3). This shows that over a ten year period the study will have cost £3.6 million and savings on information alone are expected to exceed £2.1 million.

Savings on investment are difficult to predict but improved quality and more extensive data will improve the robustness of projects and allow options such as beach recharge to be developed. It is expected that investment savings alone could pay for the whole of the SDMS cost several times over. All of these savings ignore the improvements to the technical and political effectiveness of the Region.

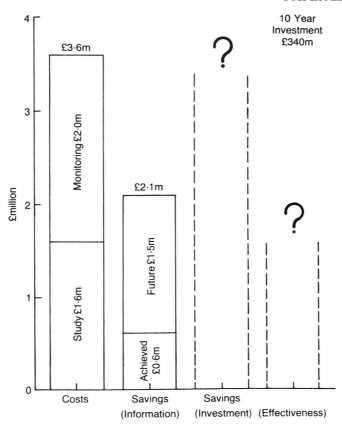

Fig. 3. Projected 10 year savings

Long term benefits

Monetary savings can justify the project in the short term but it is the longer term intangible benefits which are likely to have the greatest pay off. Some of these benefits are likely to be

(a) early information on and perhaps early warning of problems associated with sea level rise
(b) moving towards "soft engineering" (working with nature rather than against)
(c) a basis for assessing changes in coastal processes and response
(d) a sound basis for environmental statements
(e) contribution to the wider aspects of coastal zone management.

CASE STUDIES

The way forward

Introduction

The database developed as part of the project is only a snapshot in time. It needs to be kept alive to produce robust design and trend data. To achieve this an annual programme of monitoring has already commenced. The GIS System forms a basis for a management strategy of the coast. Such strategic management approach is fundamental to Shoreline Management and key to the wider aspects of coastal zone management strategies are important. Of equal importance is the need to invest in Research and Development and apply the knowledge gained particularly in the area of "soft engineering". The future is therefore about making effective use of what has been achieved and to manage the application and development.

The aim of the monitoring programme

The aim of the monitoring programme is to provide data in a structured, standardised and quality controlled manner. The data to be collected include (ref. 5)

Forcing	**Response**
(Those components acting on the coastline to produce changes)	(The response to the forcing components in terms of shoreline changes)
Winds	Aerial survey
Tidal prism in estuaries	Hydrographic survey
Water levels	Beach survey
	Inspections

The data to be collected have been carefully reviewed and refined to ensure that all data will have a specific use. The timing and frequency of collection of the response data have been fixed such that trends can be identified.

Monitoring programme

A programme of monitoring representing an investment of £0.26 million per year has commenced. Other North Sea countries have similar coastal processes and defence needs to the Region and a survey (ref. 5) has been carried out of their annual monitoring programmes.

(a) *Denmark* - a programme of annual surveying, aerial photography, water level and wave measurement for 500 km of North Sea coast costing approx £0.33 million per annum

(b) *Germany* - occasional surveys of the entire North Sea coast and more specific annual programme for the East Frisian Islands; the latter alone have a seaboard of some 200 km and the expenditure is approx £0.32 million per annum

(c) *Netherlands* - comprehensive monitoring programme of entire 450 km of coast which costs £1.60 million per annum

(d) *Belgium* - bathymetric and aerial surveys in the spring and autumn on 65 km of coast at a cost of £1.16 million per annum.

The annual costs per km are compared in Table 1.

If the estuarial length is included then the cost in the Anglian region falls to £210/km. These figures indicate that the investment in the monitoring activity proposed for the Region is generally much lower per km than those for the rest of Europe.

The need for monitoring

Monitoring is the lifeblood of the GIS system and ultimately the management strategy. The GIS system holds the available information and data and enables it to be displayed and analysed with reference to its spatial or geographic characteristics. The updating of information and data on the GIS is essential in order to carry out meaningful analysis, develop understanding of processes and monitor the performance of flood defences. This will be particularly important for soft engineering defences which require regular monitoring to establish recharge and recycling needs.

Monitoring will also allow patterns of change to be identified and therefore it is likely to provide an early indication of the impacts of sea level rise. As time goes on, a consistent and reliable data set will be available to provide sound long-term analysis and enable predictive models to be developed of how the coast will react to specific types or series of events. This will further enhance design capabilities and provide significant cost savings in avoiding both over and under design.

Table 1

	Length of open coast (km)	Annual expenditure (£m)	Annual cost/km of open coast £
NRA Anglian Region	375*	0.25	670
Denmark	500	0.33	660
Germany	200	0.32	1,600
Netherlands	450	1.60	3,550
Belgium	65	1.16	17,860

*Open coast only. (1200 km including estuary and sea defences.)

CASE STUDIES

Monitoring enables the coast to be managed in sympathy with the natural coastal process and to develop effective flood protection to reduce flood risks. This is good for the environment, for those living in flood risk areas and for the taxpayer who has to foot the bill.

Implementation of monitoring programme

The programme is managed from the Regional Office at Peterborough. However, an important concept of the management approach is the involvement of local operations personnel in the collection of data and understanding of the purpose for the data. Special software has been written to assist operations personnel in the collection and quality control of data and to provide useful data for day to day management of the coastline. Each District Office has been equipped with the necesssry software and hardware to view the data and to provide digitised data for input into the central GIS systems. Since completion of the study early in 1991 much work has been done on the monitoring programme and the letting of the necessary contracts.

District Councils with coast protection responsibilities have also been actively involved in this monitoring programme and the collation of data. District Councils with major lengths of coast protection are equipping themselves with software and hardware similar to the Anglian district offices. With monitoring now underway it is intended to move forward and develop the strategic part of the GIS and its broader role and its contribution to aspects of coastal zone management.

Towards coastal zone management

The GIS system provides, for the first time, a sound basis for the development of a management strategy. The management strategy developed and the information being collated will provide a vehicle for the wider discussions necessary on policy guidelines and management response. Development of this area will commence in 1992 and will involve wide dissemination and discussion with local authorities and all those involved in the coastal zone.

Application

Application of the information and knowledge into specific project and strategic plans is an ongoing process and will bring many benefits as already mentioned. Soft engineering depends entirely on understanding processes and monitoring performance. Without this knowledge such soft engineering or "working with nature" approaches will not be possible. The monitoring of performance of the soft engineering solutions developed is integrated with the annual monitoring programme and fed into the central GIS. Monitoring programmes have already been developed and commenced for projects in the region

(a) Heacham Hunstanton beach recharge (completed in 1991)
(b) Clacton beach recharge (completed in 1987)
(c) Happisburgh Winterton (due for commencement in 1992)
(d) Lincolnshire Strategic Works (due for commencement in 1993)

or projects represent some £100 million of investment mostly on beach recharge.

Research and development
The SDMS has already provided insight into several research needs. The NRA approach is to feed research requirements into a national procedure system and to develop research on a structural basis. The development of process knowledge and the shoreline management approach will provide key areas for structured research for the future.

Conclusions

The need for strategic investment plans for such an extensive and varied coast has prompted a unique regional study. From the outset it was recognised that any strategic approach required a fuller understanding of the coastal processes and in particular the causes of foreshore recession and lowering. The Study was completed in April 1991 at a cost of £1.65 million.

The components of the Study have been diverse and wide ranging. In the early stages of the Study many of the initiatives were exploratory in nature, seeking to evaluate various techniques. These have been developed through extensive field work, some sophisticated and highly advanced modelling work, and GIS to control and manipulate the data. In order to advance and evolve the strategy the database must now be kept up to date and for this reason a regional monitoring programme has been adopted.

The information now available provides a comprehensive platform upon which to develop a regional strategy. This will steadily improve as the monitoring programme proceeds and supplies a continuing feedback on coastal change.

Further work is planned within the Research and Development programme of the NRA, covering a range of subjects including saltings, wave forms, tidal prisms and extreme water levels. In addition the liaison with planning authorities, the maritime councils, conservation bodies and other interest groups will play an essential part in formulating future policy.

All these activities combine to make the Region more effective and efficient in its role of protecting people and property against flooding from rivers and the sea. Additionally the Region is now well equipped to make a positive contribution to the wider needs of coastal zone management.

Acknowledgements

This paper has been prepared with the invaluable support of colleagues in the Engineering Department of the Anglian Region and the Coastal Department of Halcrows, to whom thanks are due.

References

1. HALCROW. Sea Defence Management Preliminary Study for Anglian Water, 1988.
2. FLEMING C.A. The Anglian Sea Defence Management Study Coastal Management Study proceedings of the conference.
3. HALCROW. Sea Defence Management Study for Anglian Region NRA, 1991.
4. HALCROW. Reports and Manuals from Sea Defence Management Study. (See Bibliography).
5. NRA and HALCROW. Future of Shoreline Management. Conference Papers, November 1991.

Bibliography

Reports
Preliminary Study.
Inception Report, September 1987.
Supplementary Studies Report, November 1988, including:
  Wave Climate (Hydraulics Research Ltd, Halcrow).
  Residual Currents (BMT, Ceemaid Ltd).
  Beach Profile Analysis (Hydraulics Research Ltd).
  Extreme Sea-Levels (BMT, Ceemaid Ltd).
  Sea-Level Change (University of East Anglia).
  Literature Review (Imperial College).
Stage II Study Report, November 1988.
Stage II Strategy Report, November 1988.
Anglian Coastal Management Atlas, 1988.
Sea Defence Management Study
GIS Review, July 1989
Field Survey Report
  Volume 1 - Bathymetric Survey, September 1990 (BMT, Ceemaid Ltd).
  Volume 2 - Geological Survey, August 1990 (British Geological Survey).
  Volume 3 - Estuary Sediment Trends, May 1990 (GeoSea).

Data Collection and Analysis Programme for the Anglian Shoreline, May 1990 (Shoreline Management Partnership).
Monitoring Guidelines
  Main Text, February 1991
  Annexe I - Satellite Data Classification, September 1990 (University of Durham).
  Annexe II - Assessment of Beach Survey Methods, January 1991 (Hydraulics Research Ltd).
Study Task Reports
  Sediment Modelling, February 1991 (Hydraulics Research Ltd, Imperial College, Halcrow).
  Tidal Circulation, February 1991 (Halcrow, UKAEA Harwell).
  Offshore Banks (plus demonstration disk), February 1991.
  Impact of Sea Level Rise, February 1991 (University of Durham).
  Impact of Climate Change, February 1991 (University of East Anglia).
  Review of Essex Saltings Programme, July 1990.
  Essex Saltings - Research Needs, September 1990.
Sea Defence Survey Report, April 1990.
Stage III Study Report, April 1991.
Management Strategy Report, April 1991.

Manuals
Shoreline Management Data Model
EMS User Manual
SANDS User Manual

23. The defences for Lincolnshire

R. S. THOMAS, Divisional Director of Coastal and River Engineering, Posford Duvivier

The paper is concerned with the sea defences between Mablethorpe and Skegness on the Lincolnshire coast. The history of flooding, the nature of land protected and development of the defences are briefly discussed. The author describes how these defences have been the subject recently of a detailed engineering, environmental and economic assessment, leading to the development of a sound strategy for the future protection of the area.

Introduction

The National Rivers Authority Anglian Region (NRA) are responsible for the 24 km of sea defences extending from Mablethorpe to Skegness in Lincolnshire (see Fig. 1). They defend an area of some 20,000 hectares of low-lying land including in excess of 15,500 residential properties and 18,000 caravans, as well as extensive agricultural, commercial industrial and service-related activities. Prior to 1953 the sea defences were revetments (of various types) and concrete seawalls, either slabwork or stepwork with wavewalls etc, many of which were severely damaged and breached during the storm surge of that year in which 41 people lost their lives. Reconstruction of the defences began immediately after the 1953 flood. Since then many of the earlier structures have required rehabilitation and the opportunity has often been taken to upgrade the defences. In the mid-1980s the reconstruction was increased in order to cope with two problems. Firstly, many of the walls built just after 1953 were suffering from old age and secondly, there was a long-term lowering of the foreshore.

Despite the immediate reconstruction programme nearing substantial completion and therefore with a totally sound sea defence in prospect (at least for the next few years) NRA were not prepared to rest on their laurels. Instead they were proposing, if possible, to capitalise on their recent investment by providing higher beach levels throughout the Mablethorpe to Skegness length as an alternative to continued piecemeal reconstruction of the sea defences as and when structural conditions and/or beach erosion dictated. Accordingly, in 1990, NRA commissioned Posford Duvivier to undertake a comprehensive study, which would be aimed at investigating the feasibility of beach raising works and which would include mathematical

CASE STUDIES

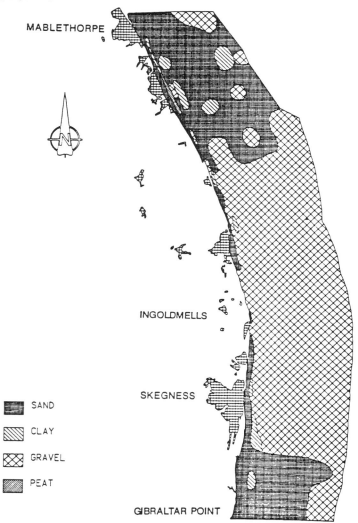

Fig. 1. Coast of Lincolnshire covered by study

modelling, include an Environment Statement, examine the financial worthwhileness and select a preferred strategy.

The coast

The study required an understanding of all the characteristics of the coast. Relevant aspects were the existing defences and beaches, winds, waves, tides, surges and currents. Whilst good basic information was available from the NRA Sea Defence Management System which formed an excellent starting point, much more detailed information was required for this study. The study concentrated on the coast between Mablethorpe and Skegness but also considered, to complete the coastal system, an area from north of Mablethorpe, extending as far south as Gibraltar Point.

The coastline

The coast generally aligns from north to south between Skegness and Ingoldmells before turning to some 25 deg west of north between Ingoldmells and Mablethorpe. There are concrete seawalls along some 19 km of the coast. Over the remainder, the defence is a reveted relic dune system. There are also around 280 timber groynes of varying length and condition. The foreshore in front of the defences is a thin mobile veneer of sand overlying a clay substratum. Currently only about a quarter of the toe of the defence remains dry at high water (spring tides).

Matters of particular environmental importance are

(a) Gibraltar Point, which lies at the southern limit of the coast and is a Site of Special Scientific Interest (SSSI) and one of Europe's most important nature reserves in terms of bird populations, which provides a range of habitats which are unique in Great Britain; internationally important populations of breeding and overwintering birds are dependent on mudflats and saltings in the area.

(b) tourism, which plays a major part in the economy of the region with large numbers of seasonal visitors and around 18,000 static residential caravans.

Existing defences

There is a large variety of different types of sea defence along this coast. A detailed inspection was carried out in the middle of 1990 to establish the details of each frontage and its structural condition. Additional information was also collected from existing drawings and documents and all information was stored within Posford Duvivier's Geographical Information System.

Beaches

The coast between Mablethorpe and Skegness has narrow, relatively steep beaches with little sand cover to the clay. The beach material varies in thickness and particle size, from a thin layer of coarse gravel after a storm to in excess of 1 m of fine sand in the summer. To the north of Mablethorpe and to the south of Skegness (as far as Gibraltar Point) there is a wide shallow beach system, with abundant fine sand which is showing long term accretion tendencies.

CASE STUDIES

Coastal processes

Damage to sea defences is rarely due to high water alone nor is it necessarily due to waves alone but occurs as a result of combinations of the two.

Hydraulics Research Limited were commissioned to undertake a study into the interdependence of wave height and water level. Data on wind, waves and water levels for Dowsing Light Vessel over the period 1978 to 1987 were used to produce frequencies of occurrence (return periods) for specific combinations of water levels and wave heights. Since reducing water depths cause changes in wave height and direction it was necessary to define conditions much closer to the shore than at Dowsing Light Vessel for use as driving forces in the coastal system. The conditions near the shore were modelled in two ways in order to provide inputs into future calculations. These were

(*a*) combinations of extreme wave height and extreme water levels; these were required for different return period storms for overtopping calculations

(*b*) annual wave conditions; these were the typical ranges of wave heights, directions and durations which could be either average or severe and were used as input data into sediment transport calculations.

Sea level rise

Long term sea level rises ranging from 5-8 mm per year were examined, which accords with NRA's present policy for the area which is 7 mm per year.

Beach behaviour

Hydraulics Research and Delft Hydraulics were commissioned to carry out studies of littoral transport.

It has already been stated that the stretches of coast to the north of Mablethorpe, from Mablethorpe to Skegness and from Skegness to Gibraltar Point, show differing characteristics. The presence and depth of sands within the north and south areas was confirmed by a geophysical survey undertaken as part of this study. The survey also confirmed the absence of sand deposits overlying the clay within the central area (see Fig. 1). Wave and tidal modelling, taking account of the nature and of the bed material, was then used to assess sediment transport.

North of Mablethorpe. Sand is feeding from the offshore bank system to the coastline, but very little feeds to the south due to nearshore tidal currents and wave refraction effects. Modelling showed that only some 130,000 m^3/annum of fine sand moves south under average annual conditions.

Mablethorpe to Skegness. Around 130,000 m^3/year enters this coastal length from the north. To the south of Mablethorpe the potential transport rate increases 2 to 3 fold but clearly because of the lack of available beach material this increased potential cannot be realised. At best, therefore, the average longshore transport towards Skegness will be 130,000 m^3/year.

South of Skegness. The offshore sand banks and in particular the shore-connected 'Skegness Middle' were considered responsible for sustaining foreshore accretion in the area south of Skegness to Gibraltar Point.

Beach erosion. The NRA and their predecessors have recorded beach profiles between Mablethorpe and north of Skegness since 1959. This formed a uniquely good data set, which was invaluable to the study. The results of a trend analysis are shown in Fig. 2 and represent the historical behaviour of the beach over the last 30 years. This shows the beach to be eroding vertically at between 0.1 cm/year and 3 cm/year. This was due to the erosion of the underlying clay rather than any significant sand loss. During storm periods the overlying sand veneer is largely removed from the upper beach area and deposited offshore to form a breaker bar system. Once exposed, or with limited sand protection, the clay is eroded. The fine eroded clay is retained in suspension and carried further offshore and effectively lost to the system.

Interaction with defences

The mechanisms of failure of a sea defence can be complex but are categorised by three fundamental modes

(*a*) erosion at the toe
(*b*) failure of the structural elements of the defence, either due to old age or impact from severe storms

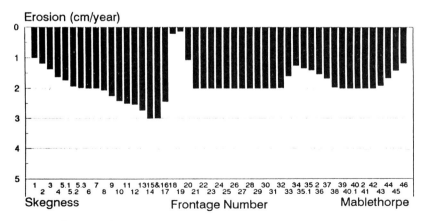

Fig. 2. Foreshore erosion rates

CASE STUDIES

(c) erosion and failure of the backslope by excessive overtopping during storms.

Toe erosion. Failure of the toe can be the result of either short term erosion due to storms, and/or long term erosion due to beach material loss. Short term erosion can be up to 2.5 m (from beach profile data). Long term erosion rates were derived by statistical analysis of the profiles and were up to 3 cm/year. These trends were extrapolated forwards for 50 years, using a mathematical model at three locations and taking account of both sand transport and clay erosion. Erosion rates were predicted to increase by up to 2 cm/year by year 50.

Residual lives of structures. One of the objectives of the survey of sea defences was to assign residual lives to the various elements of the sea defences and thereby establish the likely programme of works necessary over the next 50 years. Account was taken of changes in wave climate brought about by predicted foreshore erosion and sea level rise and the increased risk of failure due to toe erosion.

Overtopping. In extreme storms waves can overtop the defence. Overtopping can be tolerated up to a certain limit, beyond which damage will occur and this may eventually result in a breach. The assignment of limits for allowable overtopping must take into account the type of backface and crest, and the extent of acceptable damage. On the basis of Hydraulic Research Report EX 924 and other published data, limits were selected to define the likely onset of damage due to overtopping. In this study the HR limits were tested against past known events, and it was considered that they represented the onset of significant damage which, without intervention, would result in failure on a subsequent lesser event (i.e. the Do Nothing condition).

However, in reality, the NRA repairs its defences when damaged, so in considering sustaining or improving the existing defences, it can be argued that the HR limits underestimate the amount of overtopping required to create a breach during a single storm event. Following careful consideration, the overtopping limits, for these cases, were revised by a factor of four up to a maximum of 200 litres/m/s (i.e. the Repair condition).

The existing defences were analysed for overtopping for various different wave and water level conditions. This allowed the derivation of the return period of the storm event which would cause overtopping in excess of the assigned overtopping limits, for both the Do Nothing and Repair conditions.

Options

The key issues in the formation of a good strategy for sea defences on this coast were the continued and, in some locations, accelerating erosion of the foreshore, sea level rise and the short residual lives of some of the structures.

In the development of a viable strategy seven alternatives were considered either singly or in various combinations.

Retreat. Nowhere was retreat considered to be a viable approach because of the catastrophic damage that would ensue, but it was assessed to provide the base line for the economic and environmental assessments.

Set back. Set back is defined as setting back the defences to a new line. The NRA had evaluated this alternative on several schemes on this coast in the past and in every case it was rejected on economic grounds.

Sea walls. This alternative is effectively a continuation of the status quo, i.e. repair and rehabilitation as the need arises.

Beach nourishment. This alternative was studied firstly as an overall strategy in its own right and also in conjunction with other alternatives such as rock groynes or offshore breakwaters.

Rock groynes. The introduction of major structures such as long rock groynes, by themselves, would lead to significant but adverse changes in the local rates of erosion and accretion because of the limited supply of beach material available at present. They could not therefore be considered as a viable approach without the inclusion of major beach nourishment.

Offshore breakwaters/reefs. Much of the comments on rock groynes above also applied to the use of offshore breakwaters and/or reefs.

Headlands. Artificial headland types of structure would not operate effectively on this coast whose orientation is already fixed by massive concrete seawalls. This alternative could only be effective by the introduction of major beach nourishment.

Of these options, seawalls and beach nourishment (including groynes and offshore breakwaters as necessary) were investigated in more detail as being the most promising.

Seawalls

Waves reflected from seawalls increase wave agitation within the foreshore area, bringing more material into suspension and increasing the potential for sediment transport and loss. Differing types of seawall have been constructed in the past, including solid and voided stepwork, smooth revetment slopes, concrete armour units and rock. Rock provides the highest degree of wave absorption and the lowest level of wave reflection. It therefore helps to reduce but not eliminate foreshore loss. Presently some 22% of the upper beaches remain dry at mean high water springs. By 2050, with foreshore lowering of some 1.5 m in some places, the corresponding figure would reduce to some 11% and hence the recreational use of the beaches would become increasingly restricted with an associated impact on the local economy.

Concerns were raised over the safety, access and aesthetics of seawalls with rock. These would be mitigated by the provision of appropriate handrailings, access steps within each groyne bay and sympathetic design.

Finally, it has to be acknowledged that the seawall option, in allowing foreshore erosion to persist, may preclude, on financial grounds, the adoption of an alternative approach, such as beach nourishment, in future years even if this proved desirable.

Three standards of defence were costed, i.e. sustaining the present defences, raising to a 100 years' standard or raising to 200 years (Table 1).

Table 1

Costs: £million	Sustain	1:100 yr	1:200 yr
Construction cost	147.9	157.5	163.6
Discounted cost	77.7	95.4	104.4

Beach nourishment

Nourishment would be directed towards providing, through artificial means, higher beach levels. The benefit of this approach would be counteracting the erosion of the clay and thus the loss of the beach, prolonging the lives of seawalls, and increasing the standard of existing defences.

The three options considered were

(a) beach nourishment: to raise the level of the beach to +4.5 mOD, after rehabilitating the seawalls where necessary, and to recharge the nourishment in the future as required

(b) nourishment with rock groynes: with the addition of rock groynes along part of the coast

(c) nourishment with breakwaters: with the addition of offshore rock breakwaters.

Beach nourishment would provide protection to the underlying clay. Erosion of the clay would therefore be significantly reduced if not eliminated. Archeological interests associated with the clay layer would be buried, but retained for future examination. Safety and access to the beach would be improved through higher beach levels and the aesthetics of the beach would be improved. The nourishment operations would largely occur in the summer period. The operation would be largely a sea based one and thus additional road traffic would be minimal. The higher beaches would increase the tourist attraction of the area. A contingent valuation survey was performed to assess the perceived benefit to beach users of the existing beach and this benefit was included in the economic evaluation within Section 4.

Typically, material would be won from a borrow area using a trailer suction dredger and then hauled to a discharge point located approximately 1 km offshore. A sinker line would run to the beach where it would discharge.

Beach nourishment alone

Extensive mathematical modelling was undertaken in conjunction with Delft Hydraulics to study the potential sand movement. The following items were considered in detail: selection of sediment size, development of a minimum profile/volume, analysis of sediment budgets, cross shore transport, longshore transport and storm losses.

From these basic factors the whole coastline was analysed and features on a global scale were studied, such as mixing of existing and new sediments, effects of short timber groynes, discontinuities in the coastline (points), and renourishment of the beaches.

Selection of sediment size. The variability of the existing foreshore can be attributed to the existing sediment size (0.12-0.2 mm). Sediment of this size is easily moved.

A reduction in the vertical variation in beach level could be achieved by using a sand with a D_{50} of 0.5 mm for nourishment as opposed to the 0.2 mm of the existing and this would have little or no ecological or recreational impact.

Minimum profile. The minimum volume requirements for the beach nourishment were calculated assuming a variety of initial beach slopes. Severe

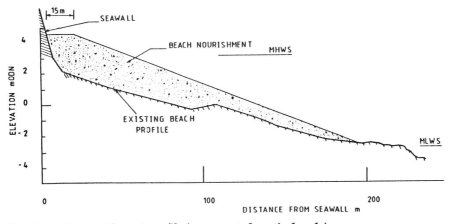

Fig. 3. Beach nourishment profile (exaggerated vertical scale)

storm modelling (up to 1:200 year event) with nourished profiles confirmed the relative stability of sediment with $D_{50} = 0.5$ mm and this material was adopted for detailed consideration. A typical cross section was developed as shown in Fig. 3.

Sediment budget. For the beach to be in equilibrium the net input of sediment has to equal the net loss. If this basic equation does not balance, then erosion or accretion takes place on the coast.

To apply this principle to the coastline as a whole would be a very coarse approach and so to get a better picture of the sediment movement the coastline was split up into 'cells', shorter lengths of around 1 km, and the equation applied to these individually. The cell has a fixed width and this was developed for each cell by the cross shore transport modelling.

Longshore transport. Longshore transport modelling showed a relatively uniform net transport rate with the only significant increase being in the vicinity of Sutton, when the shelter of the nearshore banks off Mablethorpe is lost. Under a severe annual wave climate the rates of longshore transport were shown to increase by around 45%. Longshore transport rates were significantly reduced by using a coarser sand.

Coastline discontinuities. The coastline has some notable and abrupt changes in orientation such as at Vickers Point. The existing coastline shows a consistent low water line across such features, a pattern which would be likely to persist after nourishment. As such, the higher beach levels would not be retained across such features. The option therefore included allowance for the continued maintenance and rehabilitation of such features as sea walls.

Renourishment. A key feature of nourishment options would be the need to compensate for longshore and offshore losses by periodically feeding new material to the nourished beach. Thus programmes of monitoring beach levels and renourishment would be required to return beaches to design levels and so maintain their effectiveness.

Nourishment plus rock groynes, or plus offshore breakwaters

The introduction of a groyne field into the nourished area would have the effect of reducing the longshore transport potential. Offshore breakwaters would provide a solution whereby the sediment driving forces could be controlled before they reached the nourished beach. In so doing the solution would also have the benefit of controlling both longshore transport and on/offshore transport.

The approach to the development of an offshore breakwater solution was similar to that adopted for the groyne option; i.e. the achievement of a uniform longshore transport rate of around $100\,000\text{ m}^3/\text{yr}$ rather than a total elimination of it.

Table 2

Costs £ million	Nourishment alone	Nourishment + groynes	Nourishment + breakwaters
Construction cost	203.4	219.8	196.0
Discounted cost	80.3	91.6	107.8

Costs of beach nourishment options

An appropriate programme of works was developed for the three alternative nourishment options. The resulting discounted costs, inflated to 1993 (19%), are given in Table 2.

Environmental consultation

As a first stage of the consultation process some 77 interested parties were invited to express their interests in the area. Following that a public consultation document was prepared covering the background to the factors influencing the study and the nature of the engineering options under consideration. This was circulated to 42 interested parties and comments were invited. This second round of consultations was supported with presentations and meetings.

Overall, beach nourishment was shown to be the preferred approach whenever an opinion was expressed. The approach was preferred on aesthetic, tourism, recreation and nature conservation grounds. The major residual impacts associated with beach nourishment were assessed as follows.

(*a*) A possible small increase in sedimentation may occur at Gibraltar Point. Investigations and consultations indicated that this was not likely to be significant although monitoring would be required. Chemical composition would also be monitored and any changes taken into consideration.

(*b*) The foreshore archaeological and geological interests would be covered by beach nourishment. It was, however, felt beneficial that the interests would be preserved, albeit under cover.

Economic assessment

The base line for the economic evaluation of all options was taken as 'Doing Nothing'. The residual damages associated with the adoption of differing seawall standards and the nourishment options (1:200 year stand-

CASE STUDIES

Table 3

Discounted damages	Damages: £ million
Do Nothing	944
Sea wall (residual damages)	
Sustain	253
1:100 year	64
1:200 year	55
Nourishment (residual damages)	
Nourishment alone	58
Nourishment + Groynes	106
Nourishment + Breakwaters	134

ard) were then assessed. Benefits were the difference between the Do Nothing damages and the residual damages associated with the solution under consideration. In addition, nourishment would have the increased benefit associated with the prevention of further loss of the beaches. The resulting residual damages associated with the proposed options were discounted at 6% p.a. to give presentday values as detailed in Table 3.

Contingent valuation of the beach

Beach nourishment would carry an additional benefit, i.e. the prevention of the further long term deterioration of the beaches. A contingent valuation survey was undertaken to assess the level of the willingness to pay to retain the amenity value of the beaches at their current level. A detailed questionnaire was prepared covering local residents, holidaymakers and day trippers and a total of 840 surveys (i.e. about 5% of beach users) were completed by interview between 22 July and 21 August. The study team calculated the percentage of users interviewed and to estimate a total annual number of beach visits.

This exercise predicted a total present value benefit of retaining the existing beach levels of £20.9 million.

Economic evaluation

Beach nourishment alone was shown to be the most economic on both benefit/cost and NPV grounds. The discounted costs could rise by some 22% (£18 million) before the benefit/cost ratio fell below that of the leading (1:100 year) seawall option.

Conclusions

The objective of the study was to determine the most appropriate long-term strategy for the provision of secure sea defences, considering environmental, technical and economic aspects.

Environmental

The over-riding concern of all parties was the commitment to provide secure defences to ensure the continued livelihood of this extensive area. Considerable concern remained over the aesthetic, access, and safety aspects of the seawalls. The nourishment approach was preferred on aesthetic, tourism, recreation and nature conservation grounds. Mitigating measures would have to be adopted to ensure the maintenance of the coastal sediment feed to Gibraltar Point.

Whilst nourishment could not be strictly classed as a sustainable development, in that it would require replenishment from offshore sources, it would provide a solution which would meet the needs of the present without compromising the ability of future generations to meet their own needs.

Technical

The seawall options would in the future become increasingly costly and require heavier elements to combat the increasing wave forces. Under the nourishment options, the existing variability in beach levels could be reduced by the introduction of coarser sand. Modelling showed that the losses sustained by nourishment alone would not justify the implementation of effective groynes or breakwaters.

Economic

The economic evaluation showed the nourishment option, without controlling structures, to be the most cost effective on benefit/cost and NPV grounds. The standard of defence of this option was 1:200 years.

Preferred strategy

Environmental, technical and economic considerations favoured the adoption of nourishment. The preferred strategy was to adopt 'nourishment alone' which included the commitment to monitoring and recharge.

Acknowledgement

The Mablethorpe to Skegness Strategic Approach Study, on which this paper is based, was commissioned by NRA, Anglian Region. The author is indebted to them for their comments on this paper and their permission to publish it.

24. The provision of marine sediments for beach recharge

DR A. J. MURRAY, The Crown Estate

Background

The Crown Estate is a settled estate which includes more than 250 000 acres of agricultural land in England, Scotland and Wales and substantial blocks of urban property, primarily in London. It also owns a large marine estate which includes about half the UK foreshore and almost all of the seabed out to the 12 mile limit and the right to exploit natural resources (excluding hydrocarbons) on the UK continental shelf.

As landlord, the Crown Estate is involved in a number of marine activities. One of the most economically important of these is marine aggregate extraction. Marine aggregates supply a significant part of the demand for sand and gravel in Britain, in 1991 accounting for about 15% of the national total used. The main use is in making concrete although there is a growing need for beach replenishment schemes. Extraction is generally carried out by commercial mineral aggregate supply companies under licence from the Crown Estate. This paper presents some basic facts about the marine aggregate industry in the UK and the procedures involved in obtaining a licence to extract marine aggregate. Because of the forecast sharp rise in demand for both beach recharge material and concreting over the next 10 years, competition for existing licensed reserves could make the price for beach use prohibitive. Alternative marine sources of supply for beach replenishment are examined in this paper.

Marine aggregate industry

The winning of aggregate from the sea has taken place since Roman times, and Crown Estate records show licences for sand collection from the foreshore from the turn of the century. The modern industry using sea-going vessels dredging at offshore locations moved out of foreshore and estuarial areas in the 1960s and had developed by the 1980s into a major industry supplying tens of millions of tonnes of sand and gravel annually for concrete production. There are currently about 10 companies holding 88 Crown Estate production licences covering 80 dredging areas shown in Fig. 1. Total UK and regional production from 1973 to 1991 are shown in Fig. 2. Current reserves of sand and gravel in licensed areas stand at about 400 million

CASE STUDIES

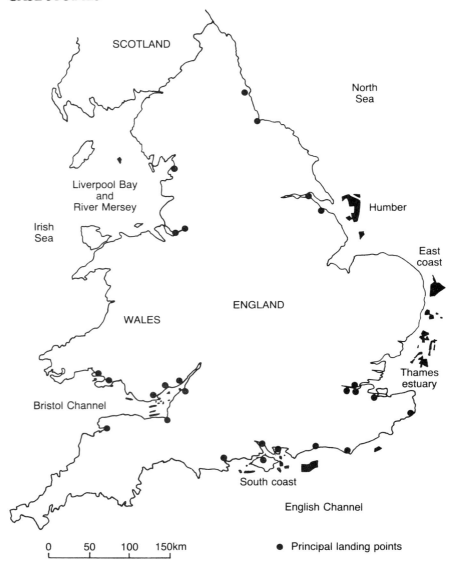

Fig. 1. Principal dredging areas

tonnes with about a further 400 million tonnes identified by prospecting surveys currently under consideration for licensing. In the South East of England, which accounts for about 75% of the marine aggregate market, licensed offshore resources stand at about 14 years at 1991 rates of extraction. The DoE have forecast an annual 4% rise in total UK aggregate requirements

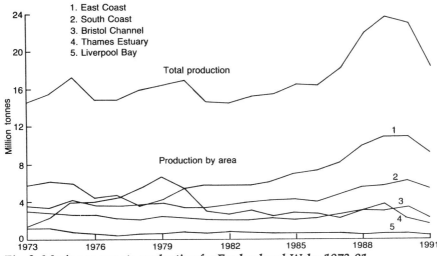

Fig. 2. *Marine aggregate production for England and Wales 1973-91*

over the next 20 years. About 50 dredging vessels land at about 100 wharves, primarily in the South East, from the Wash to the Isle of Wight. The area of seabed currently licensed is 1677 km² which represents less than 0.5% of the UK continental shelf. One km² of seabed dredged to a depth of one metre will yield 1.5 million tonnes of sand or 2 million tonnes of gravel.

Marine aggregate licensing

It is Government Policy (*DoE Marine Planning Guidance Note 6*) to encourage the use of marine aggregates wherever this is possible without introducing a risk of coastal erosion or unacceptable damage to sea fisheries and the marine environment. Generally there is no statutory planning system for the dredging of minerals from the seabed, unless the area is within the administrative boundary of a county. A licence, however, is required from the Crown Estate both to prospect for and extract minerals, including sand and gravel, from any part of the Crown Estate within Territorial Waters or the UK continental shelf. The two types of licence (prospecting and production) issued by the Crown Estate and the procedure for determining marine aggregate production licence applications, known as the Government View Procedure, are described in the following paragraphs.

Prospecting licence

Prospecting, as the name implies, is undertaken to find out about the seabed sediment in a particular area. The Crown Estate issues prospecting licences following consultation with the Ministry of Agriculture, Fisheries and Food to minimise interference with fishing activity. Modern exploration techniques include the use of seismic and sediment sampling equipment which causes minimal disturbance to the seabed. Prospecting licences are short term, usually for one year and closely defined, both in area and activity permitted. They carry no rights of extraction. If, after prospecting, an applicant requests an extraction licence, it must go through the full Government View Procedure.

Production licence

A production licence is essentially a contract in civil law between the Crown Estate as landowner and the licensee. The licence stipulates the exact area in which extraction is allowed, the amount of material which can be extracted and the commercial terms. Any requirements specified by the Government View, including conditions to protect the environment, the adjacent coastlines, fisheries and safeguard navigation, are incorporated. Control of extraction licences relies on annual audits by the Crown Estate to ensure that there is a transparent trail between the ships extracting the material and the aggregate leaving the wharf. All allegations of out of area dredging are rigorously followed up and, if confirmed, appropriate action is taken. All vessels using Crown Estate extraction licences will have to be fitted with a fully operational Electronic Monitoring System (EMS) by 1 January 1993. The EMS will automatically log date, time and dredging status and all records will be security coded and routinely checked.

Government View Procedure

The procedure for determining marine aggregate production applications is non-statutory. It is known as the Government View Procedure and is administered by the Minerals Division of the DoE, the Welsh Office or the Scottish Office. The procedure is essentially an extended consultative process which follows the principles of land-based planning arrangements. All bodies which have an interest in the Coastal Zone and the seabed, for example Fishery Departments, the Department of Transport, the Department of Energy, Coast Protection Authorities and heritage and conservation bodies, are consulted. All applications for production licences are also advertised in the local press and fishing trade press. The Government View Procedure was revised in 1989 and incorporates the requirements of the EC Directive on Environmental Assessment. The Crown Estate's role is to bring together all the relevant information gathered during these consultations. The production application, modified where appropriate in the light of comments received and discussions with the applicant, is then put to the relevant Government Department which must formulate the Government View. The Crown Estate will not issue a production licence without a positive Government View. It acts as a landowner and does not perform any quasi-planning function.

Material for beaches

In recent years soft coast protection has become increasingly considered as an effective solution to coastal erosion whilst at the same time being amenity friendly. The environmental arguments are probably more finely balanced. Although there will usually be a number of possible options for the supply of material for beach recharge schemes, the quantities involved, transport considerations and the basic nature of the material make marine sediments the favoured option in most cases.

The rest of this paper deals with the main factors which influence the use of marine sediments for beach recharge. These factors, which are all closely linked, can be considered under three main headings: availability/location, research and cost.

Availability/location

Up to now marine sediments for beach recharge schemes have mostly come from existing licensed areas. The reasons for this have been twofold: first, the material could usually be supplied on demand and, second, the specification is high and readily meets engineering and amenity criteria. With the much larger volumes proposed for future schemes it is very doubtful that the existing licence holders will be able to meet this demand, particularly in the South and East where the use of marine aggregate for construction is greatest. The alternative is to locate new sources and there are two basic options

> (a) to find material of construction quality in areas which are not available to the traditional aggregate dredging industry, e.g. in shallower water depths
> (b) to use material which is not suitable for the construction industry, e.g. too fine or contaminated with shell, lignite, clay or chalk.

Option (a)

One of the main factors which has to be considered in any application to extract marine sand and gravel is the possible impact on the adjacent coastline. All applications are examined to establish

> (a) whether the area of dredging is far enough offshore so that beach draw-down into the deepened area will not take place

CASE STUDIES

(b) whether the dredging will interrupt the supply of material to adjacent beaches

(c) whether the dredging will reduce banks which provide protection to the coast by absorbing wave energy

(d) whether the dredging will cause, or modify, refraction of waves and thus lead to significant changes in the wave pattern.

If the investigations demonstrate an unavoidable risk of coast erosion the application will be rejected without further consideration. Under the present Government View Procedure the key to further progress for an application on coast protection terms is 'nil effect'- a criterion based on the assumption that the material is being taken completely out of the marine environment, e.g. concrete production or land reclamation. If, however, the material is taken for coast protect purposes a more logical approach might be to look at 'net benefit' rather than 'nil effect'. Such an approach would mean a change to the Government View Procedure and need the full support of the administrative department and the bodies consulted under the Government View Procedure with statutory responsibilities for sea defence and flood protection, essentially MAFF, the NRA and the local coast protection authorities. Whilst such a departure from the established procedure would need to be explored very thoroughly with full consultation with all interested parties there do not appear to be any insuperable problems if the Departments concerned agree this is an appropriate way forward.

Option (b)

Marine aggregate for concreting purposes has a high specification, particularly with regard to contaminant such as silt, chalk, shell, clay or lignite. For beach recharge purposes the presence of such contaminants would not necessarily preclude its use even at relatively high concentrations particularly if the beach did not have a high amenity value. The use of dissimilar material such as sand on a predominantly shingle beach could also be considered as a way of reducing competition for relatively scarce resources.

The following section on research gives a brief description of the initial findings of a study by the Construction Industry Research and Information Association (CIRIA) into these two options.

Research

CIRIA has been commissioned by the NRA and the Crown Estate to undertake a study on artificial beach recharge. The study is in two stages: the overall objective of stage one was to investigate the potential for the use of non-aggregate marine materials for artificial beach recharge. The report will identify stretches of UK coastline where beach recharge may be an option and, where possible, the locations and availability of non-aggregate marine materials and their properties. The report will also set out the parameters governing the performance of non-aggregate marine materials and identify the scope for less rigid specifications which would accommodate the wider use of such materials. The major factors influencing the cost of supplying and placing this material on the beach will be discussed and an outline cost model developed. The overall conclusion to the stage 1 report will be that considerable scope exists for the use of non-aggregate materials for beach recharge. Although existing prospecting information does not extend fully into the near shore zone, there are sufficient indications to suggest that a similar distribution of materials lies within the 18 metre contour to that lying outside.

The overall objective of stage 2 is to develop practical guidelines for the design engineer for the supply, placing and subsequent monitoring of the performance of such materials. At the time of writing this paper the detailed format for stage 2 has not been finalised and sources of funding have not yet been agreed.

Financial implications

The current arrangements under which existing licence holders supply an agreed quantity of material of a given specification can result in some very expensive beaches, essentially because of the economics of supply and demand with a generally limited resource. An alternative approach would be for a coastal defence authority to prospect for suitable material and to obtain a production licence from the Crown Estate with an agreed royalty rate following a positive Government View. The supply of material for beach recharge schemes could then be tendered on the basis of a known location, quantity and specification. Competition from alternative users and the resulting commercial pressures on costs would be significantly reduced. There is also a major potential benefit for the engineers in having precise information about the quality and quantity of material at an early stage in the design of the coast protection/flood defence scheme.

The major initial expense for any organisation wishing to obtain an extraction licence is likely to be the cost of the prospecting survey and any work required to obtain a positive Government View on reserves which have been identified.

Prospecting for marine material has always been considered a high risk activity financially, mainly because most prospecting surveys do not identify acceptable material in commercial quantities and several surveys may be necessary. Without exception the main item of cost in prospecting is the charter of the survey vessel, which in simple terms is made up of the daily rate and the number of days required to find a suitable reserve (quality and quantity). As a very general guideline the cost of a prospecting survey is likely to be in the region of £50,000-100,000.

The review of the Government View Procedure in 1988 incorporated the requirements of the EC Directive on Environmental Assessment and in recent months the responsible Government Departments have made it clear that they expect to see a full Environmental Assessment of any application for an extraction licence submitted for a formal Government View. The Crown Estate requires any applicant for an extraction licence to submit a proposal which fully meets the DoE's basic requirements.

Once an extraction licence has been granted pre and post dredging monitoring requirements may also incur significant expense. Bathymetric and sidescan sonar surveys have been required most frequently in the past but biological and geological studies are also likely to be required in the future which may well follow on from studies undertaken as part of the Environmental Assessment. The more sensitive the extraction area the greater the monitoring requirement is likely to be.

Conclusions

Marine aggregates have been used successfully for beach recharge schemes in the past. The need for such schemes is set to increase substantially over the next ten to twenty years and the traditional sources are unlikely to be able to supply the demand completely. Competition for scarce resources will almost inevitably result in increased prices. Beach recharge materials should be available which do not compete directly with the construction market. Detailed information on the location, quality and quantity is not yet readily available. One option is for coast defence authorities to prospect for the material and obtain an extraction licence. Their efforts might best be concentrated in areas where licences for construction material would not be forthcoming, for example because of the potential impact on the adjacent coastline (the net benefit approach) or looking for material which would not be of interest to the construction industry, for example because of contamination with shell or chalk. If the quantities required justify it, there are likely to be clear financial, technical and management benefits for the Coast Protection or Flood Defence Authority which has its own extraction licence.

Discussion

It was suggested that a wider approach to coastal management enabled engineers to consider a wider range of options. It was announced that MAFF has set up a committee to review the collection of wave data. A question was asked as to whether or not a move could be made to initiate regional and local Coastal Zone Management (CZM) plans in the Anglian Region of the NRA. A strategy plan was due to be produced in 1992-93 and catchment management plans incorporating a strategy would be produced.

The proposed solution for the Lincolnshire coast was sand beach nourishment without control structures. The seabed was relatively flat offshore and the rate of littoral drift was low. An EIA was being prepared for the offshore borrow area before a dredging licence was obtained. Pumped extraction drains were not considered to be appropriate due to high clay levels below the beach. Tracers as such were not proposed but a four-year monitoring programme was planned.

It was reported that the Japanese had built about 4800 breakwaters which it was alleged were costly to maintain and unsatisfactory in operation, and that in consequence they were turning to headland control. Another view expressed was that it was misguided to generalise on options as each case had to be considered on its merits. Breakwaters had been successful at Monte Carlo, for instance. Increasingly beaches were being designed to suit available material for beach nourishment.

Comparisons were made with coastal management in Denmark, Germany and Holland. It was suggested that Denmark might be ahead of the UK in foreshore management practices but not necessarily in solving interdisciplinary conflicts. Pumped extraction drainage lines had been inspected in Denmark and a trial in Norfolk was currently planned. While capital costs for this were low the running costs were high.

25. Difficult decisions

K. J. RIDDELL, BSc, ACGI, MICE, MIWEM, Director, Babtie Dobbie

Phycisists like to think that all you have to do is say these are the conditions, now what happens next? - Richard R. Feynman

Introduction
The coastal zone is, by definition, an area of conflict and is, in fact, the physical world's solution to the problem of transition between a marine and terrestrial environment. This solution is, however, a dynamic one which modern society finds difficult to accommodate. In early times, human settlement patterns were, to a large extent, determined by climatic changes and earth movements. 'Permanent' occupation with significant infrastructure development was restricted to higher ground with solid foundations. Coastal settlements were the province of the fishermen who, apart from the occasional dramatic miscalculation, accepted the trade-off between flooding and erosion risk and the maintenance of their livelihood.

Our more recent predecessors had advanced technology to the point at which it was thought that civil engineering construction could artificially stabilize the continually changing coastline, such that urban development could proceed unhindered. Our perception has now grown such that we can see the limitations and problems of such an approach. Not only are we attempting to adopt a much wider perspective as regards the regional implications of addressing specific flooding and erosion problems, but we are trying to put such issues into their appropriate social and environmental context. It is this re-definition of the problem that makes the title of this paper occur more frequently.

The physical processes at work on the coast are many and varied. It is of particular importance to realise the vast difference in time-scales that relate to these processes. This time-scale variation is directly correlated with the geographical area one must consider. A single wave may have a time-scale of five or six seconds, the tidal cycle 12 hours, ecological disturbance one year, climatic change and human developments several years, geo-morphological reformations centuries, and tectonic movements millennia. This ratio of 10^{10} is, nevertheless, one which is being encompassed by modern approaches to coastal management.

There are as many parties interested in the coastline as there are processes under way. Fig. 1 shows some of these interests and these will all have a bias

CONCLUSIONS

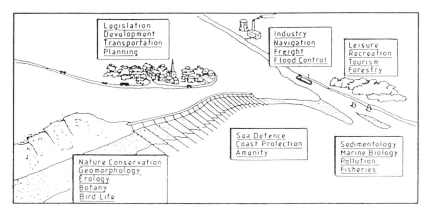

Fig. 1. Factors in the coastal management system

in favour of a particular region in the time-scale distribution. In general, most people are concerned with events ranging from the near-instantaneous to their own lifetime with, perhaps, some extension to that of their chiidren. This clearly indicates a preference for a civil engineering design life of 50 to 60 years, except for the most prestigious or sensitive projects. Inevitably, confiicts of interest can and do arise and most decision-making is concerned with the resolution of these conflicts.

The political arena is the right and proper place to interpret public opinion, set priorities and establish a framework within which defined aims and objectives can best be met. This framework will, to a large extent, revolve around financial control. How often have we heard of plans being compromised due to lack of funds? This is not an aberrent and arbitrary restriction but a reflection of public desires through the medium of how much we are prepared to pay. This is the essential background to the whole of our decision making.

Environmental issues

A holistic view of coastal management necessarily means that environmental concerns are no longer a side-issue to the consideration of a scheme to alleviate flooding, prevent erosion or attract tourists. Any coastal study is now seen as a truly environmental appreciation involving physical, socio-economic and ecological factors. There are a large number of SSSIs designated around our coastline and a significant proportion of the coast is now designated 'Heritage Coast'. These are areas where conservation, in particular, is an important factor in the decision making process. However, even in these areas, there is a move away from an approach which regards conservation as meaning preservation of the status quo. It has been realised that by attempting to preserve the status quo we may be merely trying to freeze a time-slice of a dynamic and perpetually changing environment. More attempts are being made to maintain a diversity of environments and to adopt management policies which can be revised in accordance with changing circumstances.

A particular example of the above is the increasing awareness of the liability of 'retreat options'. The deliberate decision to no longer maintain sea defences or coast protection works involves an acceptance of dramatic environmental change, albeit accomplished by natural processes. Land which is at present pasture, used for grazing and drained by fresh water drainage ditches, may revert to salt marsh affected by tidal activity and colonised by quite different species from those existing at present. Dependent upon a more global perception of the availability and importance of these habitats, such a change may well be regarded as an environmental benefit rather than a drawback. It is, therefore, important to acknowledge that any management policy will have a number of environmental implications, some positive and some negative. It is to be hoped that an adopted policy, when assessed overall, is not merely neutral but yields a positive environmental benefit.

There are more general (and global) issues which have an effect on coastal works as they do on any other area of human activity. One of these, which has generated some debate recently, is the use of tropical hardwoods in the construction of marine works. The desecration of the rain forests, as a global issue, is well known and there are opposing viewpoints as regards the continued use of this material. The supporters of its continued use argue that the consumption of specific hardwoods for this purpose is so low as to be well within the limits of acceptability for the re-stocking of managed forests. Opponents argue that not only are the forests not managed but that the consequent destruction of tropical timber is orders of magnitude higher than the consumption of usable timber. This is due to the problems of selection and access to the usable material. Others, observing this debate with interest, argue that, at the present time, regardless of who is proved right in this debate, engineers should be seen to be doing all they can to minimise any activity which is perceived by the public to be potentially harmful on a global scale. This is a difficult decision that will have to be made in the near future.

The use of our coastline for amenity and recreation purposes is so fundamental to the British view of the coast that it has long been taken for granted, but it is only recently that it has entered into our thinking as a positive benefit of coastal management. Recent research by the Flood Hazard Research Unit at Middlesex Polytechnic has moved a long way towards quantifying this as a national resource. Private developers have long been aware of the value of a 'beach'. Over most of Southern Europe, the most attractive beaches are in private hands and sun worshippers or water sports enthusiasts pay dearly

CONCLUSIONS

for the right to use them. In the USA, virtually the whole of Miami Beach was imported, placed and paid for by private business, well aware that the presence of these golden sands would guarantee their livelihood. This human aspect of environmental considerations is one that is too often omitted from the thinking of engineers and environmentalists, but is often top of the list for Town and County Planners.

Social standards, cost-benefit and funding

In this country, social expectations are not necessarily the same as the rights of an individual. Most people expect not to be flooded and expect not to lose their home due to erosion. These may appear to be not unreasonable expectations. However, these risks do exist and there is no authority obliged to prevent such events occurring. The Land Drainage Acts and Coast Protection Act give permissive powers only to the Maritime Local Authorities and the National Rivers Authority. These powers are disposed of in the light of financial constraints and the appraisal of individual situations by the method of cost-benefit analysis. It is only those schemes that achieve a benefit to cost ratio greater than unity that will proceed.

In general, sea defence and coast protection schemes depend, for their implementation, on significant support from Central Government by way of Grant Aid. Most schemes are therefore competing for a share of the funds available and they do this on the basis of benefit/cost assessment which is carried out on an increasingly standardised and rigorous basis, intended to represent the benefits to the Nation rather than potentially transferable benefits which may accrue to a particular local community. Most schemes, therefore, achieve a benefit/cost ratio much greater than the minimum and, in themselves, are optimised to achieve the most favourable economic performance. The application of discounted cash flow techniques using realistic rates of interest also means that this 'optimum' solution should be geared towards low capital cost, high maintenance schemes. However, at present, Central Government grant aid is only available towards capital works.

Due to the inherent limitations of applying solely economic criteria, the various factors that, at present, cannot be given a monetary value are known as 'intangibles' and an attempt is made to give them appropriate consideration at all stages of scheme selection, development and implementation. Factors which still come into this category for most schemes include

(*a*) stress factors affecting health problems amenity and leisure benefits
(*b*) loss/gain of natural habitats
(*c*) effects on migration patterns
(*d*) effects on geomorphological development constraints on future planning and development satisfaction of public expectations.

Table 1. Changes in National Coast Protection Policy (1986 data after Ricketts[2])

COAST EROSION MANAGEMENT COMPONENTS	U.S.A. 1986	U.K. 1986	U.K. 1992
A. INPUT OF PROFESSIONAL INFORMATION AND ANALYSIS			
1. Development of management policies	Included	Not included	Included
2. Design of engineered structures	Included	Included	Included
3. Analysis of bio-physical system:			
a) geological	Included	Partially	Included
b) geomorphological	Included	Partially	Included
c) ecological	Included	Not included	Included
d) environmental impact	Included	Not included	Included
4. Analysis of socio-economic system:			
a) comprehensive cost-benefit analysis	Included	Partially	Not included
b) planning responses (i.e. land use)	Included	Partially	Included
c) socio-economic impact	Included	Partially	Included
5. General scientific and technical expertise	Included	Included	Included
B. POLICY GOALS AND OBJECTIVES			
1. Definition of policy goals for coast erosion	Included	Included	Included
2. Definition of policy goals and objectives for CZM	Included	Not included	Not included
3. Definition of regional policy goals and objectives	Included	Included	Included
4. Re-evaluation and modification of goals and objectives	Partially	Not included	Included
C. POLICY MEANS			
1. Consideration of a wide range of policy means	Partially	Not included	Partially
2. Does the consideration of policy means include the following alternatives?			
a) constructional responses (e.g. groynes)	Included	Included	Included
b) non-constructional responses (e.g. beach nourishment)	Included	Partially	Included
c) non-structural responses (e.g. land-use planning)	Included	Partially	Included
3. Can the policy means be implemented effectively?	Included	Included	Included
D. PUBLIC INVOLVEMENT			
1. Effective provision of information	Included	Included	Included
2. Education and interpretation programmes	Included	Not included	Not included
3. Representation through elected local officials	Included	Included	Included
4. Direct involvement through:			
a) public meetings and fora	Included	Partially	Included
b) referenda and other forms of direct participation	Included	Not included	Not included
c) liaison between affected owners and decision makers	Included	Partially	Included
E. POST-IMPLEMENTATION RE-ASSESSMENT AND MONITORING			
1. Policy goals and objectives	Included	Not included	Not included
2. Policy means and measures	Included	Partially	Included
3. Responses in the bio-physical system	Included	Included	Included
4. Responses in the socio-economic system	Included	Not included	Not included

Legend: ■ Included in National Policy · ▨ Partially included in Nat. Policy · ▩ Not included in Nat. Policy

Public expectations are, to a certain extent, satisfied with regard to sea defence schemes by the adoption of desirable 'standards of protection' or 'levels of service' for new works. This situation does not appear to exist for coast protection schemes where risk assessment is at a far less advanced stage. This question of public expectations leads us to the important topic of public information in general. The wide availability of accurate flood and erosion risk mapping would undoubtedly bring market forces to bear not only on future development plans but also on current property prices and insurance premiums. This area is fraught with difficulties, but it is not

CONCLUSIONS

inconceivable that market forces could meld with coastal zone management such that public expectations were closer to actual results.

Management systems

The preceding paragraphs have attempted to illustrate something of the diverse range of issues affecting the coastal zone and part of the mechanism which supports coast protection and sea defence schemes but also has a direct effect of its own. These factors combine within the coastal management system in a more or less harmonious fashion, and the difficult decisions of the title will be necessitated with a frequency dependent on how well the system deals with the various conflicts. Many comparisons have been made between the systems in place in the USA and the UK and it is not the purpose of this paper to get enmeshed in that debate. Nevertheless, these systems do have common desirable objectives and Table 1 shows a comparison (after Ricketts, 1986) between the UK and USA. The various categories in the table include items covering comprehensive professional inputs, definition of policy aims, consideration of the widest range of options, degree of public involvement and post-project monitoring.

It can be seen that, in 1986, the comparison does not appear to be favourable to the UK. Our one saving grace appears to be that, despite all the drawbacks, we can effectively implement whatever policy we do have. Since that date we have seen

(*a*) the introduction of stricter environmental controls through SI 1217

(*b*) the formation of the National Rivers Authority

(*c*) the bringing-together of the administration of funding for coast protection and sea defence under a single agency

(*d*) the formation of various Coastal Groups to encourage and represent regional interests

(*e*) the adoption by MAFF of a wider view of coastal affairs and the need for grant-aid assistance

(*f*) the continued development of modelling techniques and geographical information systems such that forecasting and forward planning are more reliable.

I believe that the situation today is more accurately represented by the final column in Table 1, but that this is no cause for complacency. There are most shortcomings in the areas of socioeconomic impact and public involvement and, in addition, an entry in the table only signifies that the activity takes place, not how adequately it is undertaken.

Various models have been proposed for a comprehensive system for coastal management. Fig. 2 shows one such framework, departing radically from the traditional approach of response to problem situations. Its key

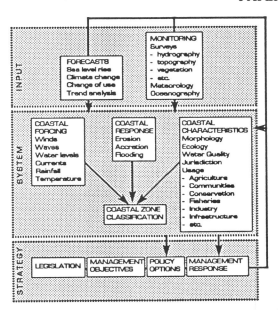

Fig. 2. Responsive Management Framework (after Townend, 1990)

elements are: the adoption of management policies rather than an assumption of capital works, the classification of coastlines and the forecasting of response in advance of actual events, and the importance of monitoring and a preparedness to change strategies in the light of improved information. This system, however, deals only with those aspects which are under the control of an implementation agency and concentrates on the scientific and technical aspects of coastal management. An attempt to set coastal management in an appropriate social setting is shown in Fig. 3. The most important player in this scenario is the public, without whose support the most enlightened policy will be destined to flounder. By bringing the public into the decision-making process to a greater extent than in the past, an obligation is placed on the professional community to educate and inform. This will become an expected and much more important part of the professional's duties in the future. Financial limitations are likely to mean that even the most careful planning will be outflanked by events and a response to public pressures and concerns will be required. The public are therefore an integral part of the event/planning/implementation cycle but can also make a significant contribution to coastal management in general, not only by making their voices heard but by their actions in the coastal environment and the political arena.

The cycle of physical processes and management actions is informed, regulated and monitored by a range of professional inputs which must

CONCLUSIONS

include fundamental research aimed at resolving some of the uncertainties which exist in forecasting and modelling techniques. The coastline is often used as the most obvious example of fractal geometry in nature. However closely one looks at it, one sees a similar degree of complexity. Such systems are often generated by very simple, but non-linear, rules which lead us inescapably to the science of chaos. In turbulent flow, weather systems and the development of the coastline, small changes can have large effects on particular parts of the system and the larger scale development will be highly dependent upon initial conditions. This may well occur within a system which exhibits a steady state or a well-defined overall trend. It is now acknowledged that we are undergoing a period of rapid climatic change which is exerting its own de-stabilizing influence on coastal processes, making forecasting a hazardous affair.

Case studies
Fairlight Cove

This area, on the south coast of England, is within an area designated an Area of Outstanding Natural Beauty for its crumbling sandstone cliffs perched on a rocky wavecut platform occasionally topped with sandy beaches formed from cliff debris. The attraction of this area is undeniable, as

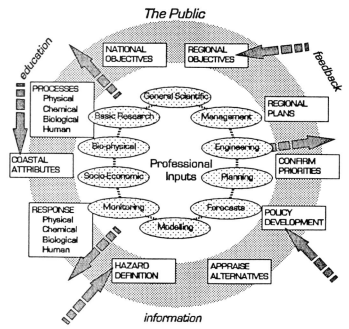

Fig. 3. Recursive Coastal Management Model

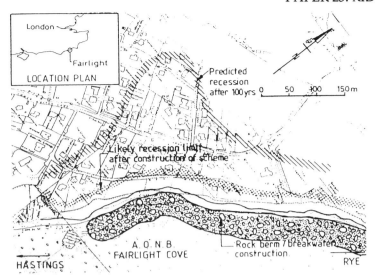

Fig. 4. Fairlight Cove Coast Protection

it has led to development taking place on the cliff-tops themselves, right up to the present time.

At present, there is very little beach material in front of the cove itself and erosion of the cliffs is dominated by the presence of a two metre thick layer of soft clay at their base which is progressively cut back by the sea, removing support from the cliffs above which collapse by massive block failure. Average recession rates in recent years have been as high as two metres per annum and at least one property has already been lost. The predominant drift direction in this area is from west to east and the present lack of beach, which leads to more rapid erosion of the soft clay, has been blamed on the construction of various sea defence and coast protection schemes west of Fairlight which retain beach material on their own frontage.

The area at risk from erosion at Fairlight is high quality residential but at a comparatively low density. Whereas sea defence schemes are assessed on a probabilistic basis, coast protection schemes are assessed along deterministic lines as the potential benefits occur progressively over a period of time rather than periodically on an almost random basis. For this reason, when the local maritime authority assessed the need or otherwise for protection in this area, not only was it seen to be very difficult to protect these cliffs but that the likely benefits did not outweigh the cost of construction. The decision was taken at this time not to proceed any further with a scheme. The local residents, to say the least, were not happy with this decision.

The residents in this area possessed between them a significant degree of expertise in a variety of fields and, over a period of time, were able to apply

CONCLUSIONS

sufficient pressure on the authority to pursue their investigations in further detail and to review their previous decision. Difficult judgements had to be made before the scheme shown in Fig. 4 commenced. These included

(a) a high degree of reliance being placed on predictions of recession lines over long periods of time, of up to 100 years, and basing the benefit assessment on these predictions

(b) the acceptance of a reduced scheme which would not eliminate erosion completely but would slow it down considerably and, perhaps, allow the cliff to stabilise its position over a long period of time; this entailed approbation of continued, but limited, property losses

(c) the curtailment of the protected length of coastline such as to optimise the benefit/cost ratio.

It is most unusual in this country for local people to 'force the issue' in this pro-active way and it remains to be seen whether the actions taken are, in fact, to everybody's satisfaction in the long term.

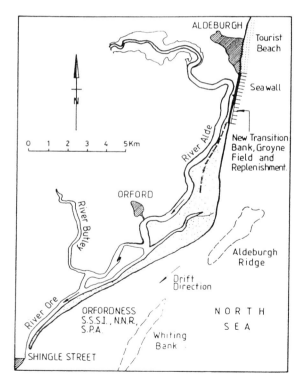

Fig. 5. Aldeburgh Sea Defence Scheme

Aldeburgh Sea Defences

The east coast of Suffolk has been receding continuously throughout the historical period. The village of Dunwich has virtually disappeared and Aldeburgh itself has lost several streets. What is now a sea-side promenade used to be the main street running almost centrally through the town. Although variable in direction, the drift is predominantly from north to south and the shingle resulting from this erosion has accumulated at Orford and produced the lengthy spit known as Orford Ness. This is rightly claimed to be one of the most important shingle formations on the coast of the British Isles and is designated a Site of Special Scientific Interest, mainly for its geomorphological characteristics. Aldeburgh Town frontage is protected by a low sea wall and extensive groyne system which has built up a wide expanse of beach, so much so, that difficulties have been experienced in launching the local life-boat. South of the town the sea wall and groynes have been less successful, failing to adequately retain material and causing the formation of a deep scour hole in the spit itself at their southern end.

By 1986 the situation had continued to deteriorate to such an extent that serious fears existed regarding the possibility of a breach at this location which would not only exacerbate local sea defence problems but may provide a new mouth for the River Alde. It can be seen from Fig. 5 that the River is perilously close to the existing coastline at this location, its course having been swept even further south in the past by the accumulation of shingle at Orford Ness. Extensive investigations were carried out including wave refraction studies, beach movement monitoring, mathematical and physical beach modelling, mathematical modelling of the branched river estuary, environmental assessment and detailed cost/benefit analysis. The results of the river modelling exercise were particularly interesting and are shown in Fig. 6. This particular section is in the main channel adjacent to Havergate Island. Even during a simulated 25 year tidal surge event, velocities in the channel are dramatically reduced and are predominantly in the reverse direction. It should be noted that this scenario assumes that both new and old river mouths remain open at this stage. The consequences of this change to the existing river regime include silting of the channels to the south of Aldeburgh with the potential consequence of closure of the old mouth and the isolation of the town of Orford from the sea, saline intrusion to a far greater distance upstream and the de-stabilization of the present meander pattern and flood-marsh area. Also, significant lengths of river bank would now be exposed to wave attack for the first time.

The preliminary cost-benefit analysis for this scheme, undertaken on a fairly rigorous, but conventional, basis, did not yield sufficient benefits to justify the promotion of a scheme to prevent a breach. The benefits derived from limited areas of residential flooding and large areas of agricultural damage. The present low perceived value of agricultural land undoubtedly

CONCLUSIONS

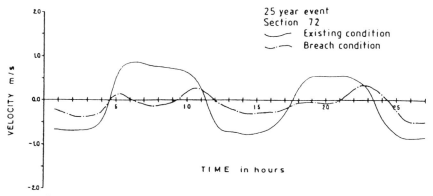

Fig. 6. *Effect of a breach on velocities in the River Alde*

contributed to the overall low benefit figure. This situation caused some degree of consternation, as the social and environmental disturbance caused by moving the mouth of a major river some seventeen kilometres along the coastline would be enormous. Further detailed investigations were initiated aimed at quantifying, in some detail, the degree of social and environmental disturbance and also to put monetary values to these effects. The scheme has now been completed, and has been successfully undertaken because the following decisions were made as regards the justification and design of the scheme

> (a) the acceptance, for the first time in this country, of environmental benefits making a material contribution to the cost-benefit assessment and, hence, the justification for the scheme
>
> (b) the acknowledgement that, despite the application of modern modelling and forecasting techniques, different policies carried with them differing degrees of risk and values of uncertainty. It was really the high uncertainty associated with a potential breach scenario, and its effects on the littoral and river regime, which swung the balance towards maintenance of the status quo
>
> (c) the design of the scheme be geared towards redressing the existing discontinuity in littoral processes and minimizing any interference with beach movements.

Chesil Beach

Chesil Beach is a natural shingle barrier, some 27 km in length, extending westwards from the village of Chiswell on the Isle of Portland to West Bay, adjacent to Bridport. It is the only major shingle beach in Britain which is essentially a simple storm beach. The beach comprises pebbles of chert and flint which are basically single sized at any particular location but increase progressively from approximately 4-5 mm in the west to 50-100 mm in the east. This grading phenomenon has been the subject of much research and speculation, but no single explanation has been accepted as factual. The beach is designated a Site of Special Scientific Interest, as are Portland Harbour Shore and the Isle of Portland, albeit for different (non- geomorphological) reasons. The following quotation from a report by the Institute of Oceanographic Science summarises the overall attitude to any works which may affect the beach.

> "Although in most respects the scientific interest of the site would be best maintained with the least man-made intention, it may be argued with some justification that an element of sea defence work is necessary to enable the very existence of Chesil Beach itself as both protective bastion and scientific entity."

Chiswell and the narrow neck of Chesil Beach, which is the Isle of Portland's only connection to the mainland, has a long history of flooding, 22 events having been recorded since the catastrophic storm of 1824 when 80 houses were damaged or destroyed and 26 people drowned. It would appear that in previous times the advantages of proximity to the shore and to their boats outweighed the disadvantages of occasional flooding to the predominantly fishing community. However, times have changed, and the local population are no longer fishermen but commuters to the mainland to whom there is no trade-off which makes periodic flooding acceptable. Severe

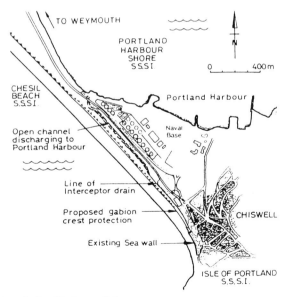

Fig. 7. Chesil Beach Sea Defence Scheme

CONCLUSIONS

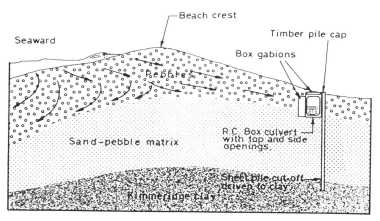

Fig. 8. Percolation through Chesil Beach

events in 1978 and 1979 precipitated investigations into the feasibility of alleviating flooding of Chiswell Village and the Weymouth Road. Detailed descriptions of the scheme and the build up to it may be found in references 6 and 7.

The main component of the scheme was an interceptor drain running for some 800 metres behind the crest of the beach. Figure 8 shows a schematized cross section of the works. The percolation flow illustrated was the main, and most frequent, cause of flooding.

Detailed ground investigations involving the sinking of 4 m square trial shafts 15 m into the beach had revealed the presence of the sandy strata shown in the figure. Large scale permeameter testing revealed that the pebble matrix was underfilled and the sand component washed out of the pebbles at high hydraulic gradients. If this were allowed to happen on site then the overall permeability of the beach would dramatically increase to that of the open large pebbles and the flooding problem exacerbated rather than relieved.

There was therefore no safe solution and conservative estimates could not be made for the various design parameters for the scheme. If the drain were to cause 'backing-up' of the flow through the beach then more of the flow would be returned in a seaward direction, thus causing increased loss of material to the sea and long term degradation of the beach. If the drain were over-capacity and drew down water levels within the beach, then the high hydraulic gradients would cause leaching of fines and progressive increases in permeability and flood flows. These considerations led to a very extensive (and therefore costly) programme of physical and mathematical modelling before confidence in the scheme could be assured. The scheme was undertaken successfully due to a combination of the following factors

PAPER 25: RIDDELL

(*a*) the close collaboration of Weymouth & Portland Borough Council, the NRA and MAFF in promoting the scheme
(*b*) the early involvement of local interests and the various environmental agencies
(*c*) the use of the most accurate and reliable methods of investigation to generate appropriate levels of confidence in all aspects of the scheme
(*d*) the use of additional planning measures and incentives to maximise the benefits accruing to the scheme and capitalise on the investment already committed.

Hunstanton/Heacham Beach recharge

The area surrounding the Wash, including the towns of King's Lynn, Snettisham and Hunstanton, suffered heavily during the east coast floods of 1953. Fifteen lives were lost in the King's Lynn area and, in response to the report of the Waverley Committee, existing walls and revetments in the area

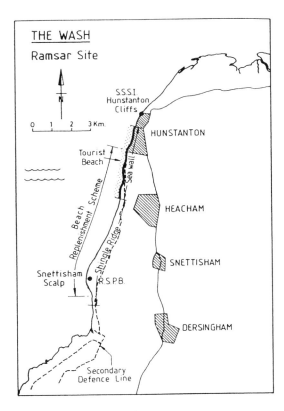

Fig. 9. Hunstanton/Heacham Beach recharge

CONCLUSIONS

were upgraded or rebuilt shortly afterwards. By 1988, continued erosion in this area had caused beach levels to drop by 2.5 m from their level when the works were constructed, and the defences were assessed to be vulnerable both to overtopping and from undermining of their toe affecting their structural stability.

Comprehensive investigations of this area revealed that, as the vast hinterland was protected by a secondary line of defence consisting of an earth bank, the direct area of benefit for any scheme was comparatively small. It was therefore necessary to seek a low cost solution to the problems presented by the deterioration to the present defences. The preferred option was for beach recharge which has resulted in a scheme costing £750 per metre run as opposed to sea-wall refurbishment and rebuilding costs of £2000-5000 per metre run but with concomitant increased maintenance costs and monitoring commitment.

The adoption of beach recharge for sea defence purposes involves cognisance of the concept of variable risk throughout the lifetime of the works dependent upon the maintenance schedule and climatic variations. Fig. 10 shows an example of the results of beach modelling necessary to determine appropriate beach recharge volumes for an acceptable combination of capital cost, risk and maintenance commitment. It can be seen that the majority of movement takes place during the first part of the storm, giving rise to thoughts of a stable profile. Three-dimensional effects probably prevent this from being the case, but one aspect that is undoubtedly a correct indication is the substantial steepening of the rear face of the beach. Since modelling techniques at the present time can only deal with single size materials, the

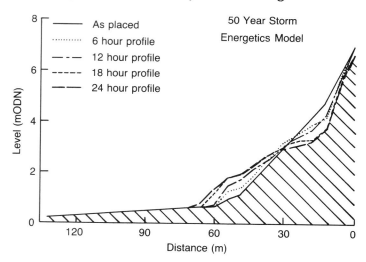

Fig. 10. Hunstanton cross-shore response

performance of a graded material can only be determined by prototype observations.

The choice of material specification on a site such as this is not easy. Reference to Fig. 9 will show the area to be surrounded by tourist beaches, a geomorphological SSSI and a bird sanctuary administered by the RSPB. The Wash itself is also a RAMSAR site. The decision was made to specify a material with a very similar grading to the indigenous beach. Despite public acceptability being one of the main reasons for this choice, it has been public reaction to the recharged beach which has brought attention to the problems caused by windblown sand and significant cliffing of a beach well used by local people and tourists.

The scheme is being comprehensively monitored with particular regard to long-term maintenance implications.

Conclusion

I have attempted in this paper to outline the framework in which today's coastal management decision making is carried out and to highlight some aspects of selected schemes which illustrate how this framework affected crucial decisions affecting our coastline. I have also speculated on the direction future coastal management practice may be taking, with an optimistic view of the continuation of current trends. At the present time there is intense debate concerning the possibility of the creation of an unitary authority for the coastline but this path is not without difficulty. The variety of local and regional interests involved necessitate representation, and the necessity to make difficult decisions is unlikely to disappear overnight. We must aim for improvements to our management systems, our methodology and our scientific knowledge to play our professional role in the Coastal Management of the future.

References

1. The development of an expert system for coastal management. Littoral, 1990, Association Eurocoast, Marseilles, July 1990.
2. Ricketts P.J. National policy and management responses to the hazard of coastal erosion in Britain and the United States. Applied Geography, 1986, 6, 197-221.
3. Townend I.H. The development of coastal management procedures. Littoral, 1990, Association Eurocoast, Marseilles, July 1990.
4. Gleick J. Chaos - making a new science. Heinemann, 1988.
5. Anglian Water, Norwich Division, Aldeburgh sea defences - Part One Report, Unpublished, 1986.

CONCLUSIONS

6. Hook B.J. and Kemble J.R. Chesil Sea Defence Scheme. Paper 1: concept, design and construction. Proc. Instn Civ. Engrs, Part 1, 1991, 90, Aug. 783-798.
7. Heijne I.S. and West G.M. Chesil Sea Defence Scheme. Paper 2: design of interceptor drain. Proc. Instn Civ. Engrs, Part 1, 1991, 90, Aug., 799-817.
8. Nairn R. B. and Riddell K.J. Numerical beach profile modelling for beachfill projects. Coastal Engineering Practice '92. American Society of Civil Engineers, California, March 1992.

26. Coastal groups

T. A. OAKES, Anglian Coastal Authorities Group

Introduction
My paper explains the formation of the coastal groups, their objectives and the way in which they are addressing the conflicts existing within the coastal zone. It suggests a way forward to produce local and regional coastal management plans.

I rely mainly on the work of the Anglian Coastal Authorities Group (ACAG) and my own authority, Waveney District Council in Suffolk.

Formation of coastal groups
During the early 1980s coastal defence was predominantly the domain of engineers. Funding had not become a major issue as expenditure on coast protection had not been reduced by Central Government. Anglian Water was limited in its expenditure too, so it had not wished to expand its role. The responsibility for grant aiding land drainage and coast protection works was split between the Ministry of Agriculture, Fisheries and Food (MAFF) and the Department of the Environment (DoE). Various environmental groups had to be consulted during the preparation of major schemes but other than those in a statutory role they were not at that time seeking a greater involvement in decision making.

Over the next few years opinions began to change and there was a call for greater consultation and participation. In 1985 MAFF assumed responsibility for the administration of schemes proposed under both the Coast Protection Act 1949 and the Land Drainage Act 1976. In this climate of change, various groups of engineers decided there would be benefit in meeting to explore issues. At this time there were no ideas or aspirations to discuss or develop regional policies.

The issues under discussion were summarised as follows.

(*a*) Funding for new schemes was becoming more difficult to obtain due to the availability of funds to meet the demand for new schemes.

(*b*) Marine pollution was becoming a very topical subject particularly with the development of EEC Regulations and the importance of Blue Flag Awards for clean beaches.

(*c*) Authorities had become aware that the increasing demand for sea dredged aggregate could create problems onshore. The Crown Estate's

CONCLUSIONS

reliance on the supportive evidence of Hydraulics Research was being questioned.

(d) There was an increasing awareness of the value of the environment. The Countryside Commission was delineating coasts of particular importance as Heritage Coasts. Environmentalists and conservationists were seeking a greater involvement in the decision making process.

(e) Traditional methods of protecting the coast by means of reinforced concrete structures, steel sheet piles and other hard solutions were being challenged in the light of a better understanding of coastal processes. It was acknowledged that there was a need to work with the sea rather than fight it and soft solutions comprising beach nourishment, armour defences and groynes were taken more seriously. There was a search for solutions using naturally occurring materials that were flexible, cost effective and more environmentally acceptable.

(f) It was accepted that work carried out by one responsible Authority was going to affect the adjacent Authorities and the concept of the coastal cell took on greater significance.

Concern about these issues led engineers in East Anglia to conclude that a Technical Forum for the exchange of technical information would provide a platform for debate, collaboration and consultation.

In May 1987 the first meeting of the Anglian Sea Defence and Coast Protection Group took place, attended by authorities from North Essex, Suffolk, North and East Norfolk, MAFF and Anglian Water (now the NRA).

In 1989 Anglian Water commissioned the Sea Defence Management Study of its coastline between the rivers Humber and Thames and the Shadow NRA was formed. For these reasons the group decided to approach other coastal authorities within these boundaries in order to determine interest in joining the group. Subsequently authorities from South Humberside, Lincolnshire, West Norfolk and South Essex joined the group which was renamed the Anglian Coastal Authorities Group (ACAG).

At the same time as ACAG was forming in East Anglia, the Standing Conference on the Problems Associated with the Coastline (SCOPAC) was developing on the South Coast. The aims and objectives of both groups were remarkably similar, but they were very different in composition.

(a) As well as the Maritime Districts and the relevant Water Authorities, SCOPAC included the respective County Councils.

(b) Each participating Authority in SCOPAC was allowed one voting elected member but other elected members and relevant officers of each authority were entitled to attend meetings.

(c) The Conference was served by an officers' working group who met separately and prepared reports, arranged meetings and speakers, and co-ordinated research.

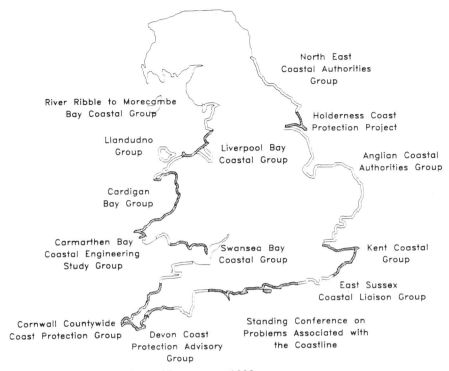

Fig. 1. Coastal Groups formed by January 1992

(d) Constituent Authorities of SCOPAC were required to make a financial contribution towards the costs of administration and other expenses.

MAFF and the Welsh Offce have supported the formation of Coastal Groups. Fig. 1 illustrates the groups in England and Wales formed by January 1992 and it is self evident that only a very small proportion of the coastline is not now covered by such groups. Most of the groups are similar in composition to ACAG.

Group objectives

During 1990 and 1991 some of the Group Chairmen met informally at the Institution of Civil Engineers and with the extension of the groups so that they covered most of the English and Welsh coastlines. MAFF and the Welsh Office formed the Coastal Defence Forum. The Terms of Reference of the Forum are shown in Appendix 1.

From these meetings it is apparent that the coastal groups do have some common objectives.

CONCLUSIONS

(a) The first is to provide an opportunity for each member to learn from the experience of others. This is the basis of the Technical Forum where new ideas and techniques are discussed.

(b) Most groups have created a library of research data and other studies undertaken by the individual organisations. This information is readily available to each member of the group for its own use.

(c) Some groups have identified the need for further studies to provide an understanding of the broad problems being experienced along their coastline. Within ACAG, the NRA study provides each member of the group with a unique opportunity to gain access to valuable information. Each authority carries out surveys of its own particular length of coastline and the results are fed into the overall database. This ensures the future maintenance of the database to which individual members will have access, either in-house, or at the NRA regional office.

(d) The last but possibly the most important objective is to produce a coastal management plan for their cell(s).

Common issues

Many of the groups have achieved the first three objectives and are now working towards the production of the management plan.

In response to a seminar held the previous year by the Institution of Civil Engineers, ACAG brought together a broad range of people interested in producing a coastal management plan at a seminar arranged in Cambridge in 1990. The organisations represented were local authorities, central government departments, statutory and non-statutory consultees and other interested professionals.

This seminar identified a series of issues affecting the coastline. These issues are considered to be common around the English and Welsh coastlines.

(a) Repairing or replacing defences on a like-for-like basis was not possible, nor necessarily desirable, in the tight financial situation within which authorities found themselves.

(b) There was a growing awareness of the value of the environment and a clear wish for environmentalists to be more involved in the decision making process. This ranged from geologists interested in a particular outcrop to those wishing to preserve a coastal habitat for wildlife. These groups were no longer prepared to see uncontrolled interference of important lengths of coastline or the loss of material to feed beaches.

(c) The policies on the approved use of land along the coastline were being questioned. The growth of "unnecessary" developments, particularly caravan sites, in the most beautiful and unspoilt areas of the coastline, was criticised strongly. These developments led to the need to protect previously unprotected coastlines.

(d) The Government had suggested changes in legislation to bring together the Land Drainage and Coast Protection Acts in the form of a Flood Protection Bill. In turn, this had led to arguments for and against the creation of one national body to oversee coastal defence grant aid and development.

(e) The development of the coastal groups had been somewhat ad hoc and there needed to be more overall guidance to co-ordinate their efforts.

(f) Off-shore dredging for aggregate was increasing, especially off the coasts of Essex, Lincolnshire and Kent. Authorities were becoming very concerned that the loss of this material would lead to erosion along their own coastlines. There was a demand for this dredging either to be monitored more closely or to be reduced.

The management issues to be balanced

The comments at the ACAG seminar reinforced the group's earlier conclusion that these conflicts had to be resolved through the introduction of a coastal management plan for the region. Three major issues were identified.

(a) The policies on land use within the coastal zone created conflict through demands for greater development to aid the local economy, tourism, recreation, residential development, and many other uses. With the increasing influx of visitors and those wishing to retire to the coast, this put greater pressure on the unspoilt coastline, particularly along the Heritage Coast and at areas of Sites of Special Scientific Interest. In face of this was the demand to return as much of the coastline as was possible to its former natural state.

(b) Justifying coastal defence schemes by including only the monetary benefits and disbenefits in the cost benefit analysis could lead to schemes unsympathetic to the environment. This could be avoided by including the value of protecting or preserving the environment in the calculations. In association with this were more pressures not to use the traditional methods of protection and construction, especially if they included the use of non-renewable tropical hardwoods. As a result, alternative materials and techniques had to be used more frequently to help conserve important areas.

(c) The absence of long term planning of schemes and expenditure led to decisions being taken on short term issues. There was little reference to the wider interests of the region.

To ensure that these conflicts are minimised or removed, the Groups have begun to develop good liaison and consultation procedures so that everyone

CONCLUSIONS

who has an interest is involved in the decisions taken on land and water use, engineering requirements and conservation.

Similarly the procedures of funding are being reviewed with MAFF to determine the extent to which the value of the environment may be taken into account at the design stage.

The demand for a sea dredged aggregate is also being reviewed. Firstly in recognition of its role as a first line of defence on the beaches and secondly, to balance the conflict of income to the State versus potential damage along the coastline.

The strategic planning between MAFF, the NRA and the Maritime District Councils on how to best protect the coastline should be continued and be built upon.

The discussion on how best to balance these issues will involve many interested parties and not just engineers. They will include the public, local authorities, several central government departments, lobbyists, and Members of Parliament. It has been proven that if one subject is certain to create interest it is the decision not to proceed with a coastal defence project.

Regional planning

ACAG has proposed that the development of regional policies should be through a "bottom up" approach. In Fig. 2, I propose a process through which the various plans could produce the local and regional policies for incorporation into a regional coastal management plan. These are explained in detail later in the paper.

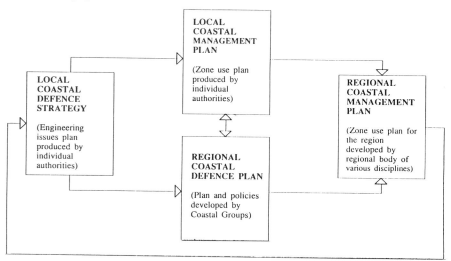

Fig. 2. Hierarchy of Coastal Management Plans

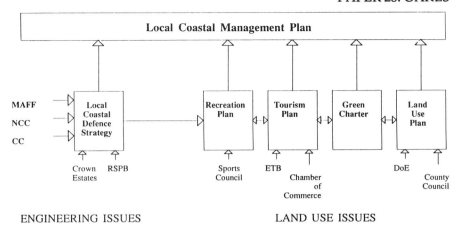

Fig. 3. Elements of Local Coastal Management Plan

Figure 3 illustrates the relationship between the various strategies which form the basis for the local coastal management plan. Most organisations have these strategies in one form or another, the difference being that they probably make no reference to similar plans of neighbouring organisations.

When complete the regional coastal management plan will serve as a general statement of policy against which all future local coastal management plans and strategies can be tested.

Local Coastal Defence Strategy

The local coastal management plan has two distinct but complementary aspects - the engineering issues and the land use issues.

The engineering issues are addressed in the local coastal protection and flood defence strategy, herefter referred to as the Local Coastal Defence Strategy. Referring again to Fig. 3, some of the external influences or factors that have to be taken into account whilst producing the strategy are denoted by the arrows. For instance, English Nature, the Countryside Commission, the Crown Estate and Royal Society for the Protection of Birds are involved. There will be other statutory and non-statutory consultees who will wish to comment on the coastal defence strategy produced by each authority and their views shoud be taken into account.

The coastal defence strategy has objectives and those suggested are

(*a*) to produce a strategy for the effective management of the coastline
(*b*) to provide an appropriate level of protection
(*c*) to identify opportunities for development and conservation
(*d*) to reinforce commitment to the coastal management plan

CONCLUSIONS

(e) to monitor coastal processes
(f) to identify requirements for further research and study.

The first objective will ensure that the strategy and its proposals play their part in protecting the coastline in accordance with national policy. This will require the production of a ten year plan.

The second objective will lead to the need to make some difficult decisions. For instance, the authority may decide that some lengths have to be protected at all reasonable costs. Alternatively, some lengths may not be protected in the future due to the unacceptable environmental damage caused by any necessary structures or defences. These are most likely to be within the lengths of the coast designated as SSSIs or Heritage Coast. Decisions on the future level of protection along the remaining lengths of the coastline could be taken on overall merit. In some of these cases there may be strong objections to further coastal defence works and in these situations the final decision would be taken after the appropriate Local Planning Inquiry. From these decisions would come an indication of those areas where, say, infill development could be allowed, or circumstances under which new developments could proceed.

Upon review the local coastal defence strategy will clearly have to support and reflect the regional strategies and plans. The NRA Sea Defence Management Study will play a vital role in the future monitoring of coastal processes in the East Anglian region. It is essential that refernce be made to it in any strategy in this region.

The last objective will generate greater co-operation between group members and establish the need for "cell-wide" studies.

Regional coastal defence plan

At an early stage local coast defence plans of the group members should be compared with others so that, through an iterative process of examination and questioning, the general and common principles to be applied across the region are identified. These major issues would form the policies of the regional coastal defence plan. Once they have been identified and agreed these policies would be referred back for inclusion in the local coastal defence plans. This would be achieved through discussion and debate within the coastal groups.

To be effective the regional coastal defence plan will need to be enforceable. It is not necessary to give the coastal groups statutory recognition but instead central government could require all proposals for grant aid to be referred to the coastal group and its regional coastal defence plan. From their work to date the Maritime District Councils and the NRA have become aware of the need to support the strategic planning process.

Local coastal management plan

With a local coastal defence strategy an authority can begin drafting a local coastal management plan. Fig. 3 illustrates some of the other land and sea use strategies necessary to complete the management plan, e.g. the recreation plan, the tourism plan, the environmental protection plan and general land use plan.

Combining these elements at the local level produces a local coastal management plan. This plan would acknowledge the local conflicts along the coastal strip and it would be used to produce policies determining the future development and management of activities within the coastal zone.

Once again this plan should not be produced in isolation from the similar plans in the group. At an early stage it should be compared with other local management plans to take into account the wider issues at the boundaries of adjoining authorities. These can then be included in the developing local plans.

Regional coastal management plan

The definition of Coastal Management Plan proposed to ACAG is that of Sorensen et al. (1984): "Coastal Zone Management typically is concerned with resolving conflicts among many coastal uses and determining the most appropriate use of coastal resources".

From the comments received at the ACAG seminar, I feel that this definition best addresses the apparent conflicts within the coastal strip.

It had been the intention of ACAG to produce, by 1994, an overall regional coastal management plan from the individual plans of the members. In its present form ACAG is well suited to produce the regional coastal defence plan but after one year's work it is apparent that the group is not suited to produce the regional coastal management plan. I believe now that it cannot and should not, expand its membership to produce the regional coastal management plan as this work involves so many different interests. For example, some of the disciplines required are planners, conservationists, environmentalists and elected members. Nonetheless, ACAG still has a major and important role to play in this task as the coastal defence strategy is a basic and essential piece of work.

If the coastal groups are not to be the driving forces behind the development of the regional coastal management plans then an alternative body has to be identified. One coastal group has some of the disciplines as observers and another also includes elected members on its committee. However, even these groups have no statutory recognition and this is a weakness.

To be effective the regional coastal management plan has to enforceable. To achieve this, the plan will need the authorisation of central government.

This is only likely to be given to a regional body containing directly elected representatives and as such this discounts existing organisations.

The closest model within East Anglia is the Standing Conference of East Anglian Local Authorities (SCEALA). This is a body for collaboration between the three county and 20 district councils in Cambridgeshire, Norfolk and Suffolk. Its main objectives are: to establish regional priorities and formulate regional planning guidance; to increase awareness of the region's opportunities and problems; to initiate action, co-ordinate effort and lobby on behalf of the region; and to promote the economic health of the region. Subscribing member authorities nominate elected councillors to attend regular conference meetings. Members of Parliament, Members of the European Parliament and the Officers of Central Government departments are invited to these meetings which are open to the public.

Clearly, SCEALA's objectives are sympathetic to the development of a regional coastal management plan. In the first instance it could develop a plan for the coastal zone of Norfolk and Suffolk but it does not cover the relevant geographical area to represent the whole of the coastal cell. Regional bodies set up on similar lines to SCEALA but addressing coastal issues could fulfil the role on a national basis.

Conclusions

To resolve the conflicts within the coastal zone and to meet future challenges there needs to be determination to develop regional coastal management plans.

The development of the regional plans will be a reiterative process of examination and debate between people from all disciplines and interests within the coastal zone.

The development of the regional coastal management plan creates the need to identify a body to act as the driving force.

The development of the regional coastal defence and local management plans should proceed in parallel.

The enforcement of the policies within the regional coastal defence and management plans will mean that some difficult and perhaps unacceptable decisions will have to be taken. In the interest of the region, a proposal from one member may not be supported for, say, a high priority for future expenditure, with the consequent loss of protection to that length of coastline. The acceptance of this principle will prove the merit or otherwise of the regional coastal management plan.

There are positive benefits to be gained from participating in and supporting the coastal groups in their development of the regional coastal defence plans. Furthermore, the groups have a vital role to play in the development of the regional coastal management plans. To continue their work requires no change in Central Government legislation but to make the defence plans more meaningful they will have to be given some recognition by central government.

Appendix 1. MAFF/Welsh Office Coastal Defence Forum terms of reference

To provide a national forum on coastal defence including sea defence and coast protection matters for government officials, representatives of Coastal Groups and others with an interest in Engalnd and Wales.

The Forum will assist authorities in the performance of their functions by

(*a*) furthering co-operation between parties with responsibility for coastal defences
(*b*) sharing experiences, data etc.
(*c*) identifying best practice
(*d*) identifying research needs and possibilities
(*e*) promoting strategic planning of coastal defence management
(*f*) identifying obstacles to progressing planned works
(*g*) keeping abreast of policy developments, R&D results and new initiatives.

The Forum will be led and supported by MAFF on behalf of MAFF and the Welsh Office.

Discussion

The view was expressed that too much change should not be attempted at one time. As the coastal cell group coverage was large there might be a need for some rationalisation of boundaries. An investigation on this was in progress. In order to realise group objectives it might be necessary to reject schemes which did not fit into the local strategy plan. It was suggested that it was not necessary for administrative boundaries to fit coastal cells.

The SCOPAC coastal group was reported to be considering extending its interests beyond shoreline protection to the wider scope of CZM but concluded that this could not be done using current powers and representation. One conclusion was to focus on the shoreline but to promote a wider approach.

Closing address

M. G. BARRETT, MBE, FICE, FRICS, Senior Consultant, Posford Duvivier, and Chairman, Conference Organising Committee

In 1982 the Institution of Civil Engineers promoted a conference at Southampton on Coastal Engineering. As its name implies this was largely orientated towards engineering. This was followed in 1989 by the Bournemouth conference which was entitled Coastal Management. The scope of the conference was broadened to include more planning and environmental issues.

Recognising the importance of co-ordinated management and the consideration of long-term issues in dealing with the coastline, on this occasion the scope of the conference has been widened even further. The question of interdisciplinary management has been concentrated on and the subject of coastal structures left to be covered by a separate conference.

This progression is an indication of the changing attitude to coastal problems in the civil engineering profession and of the direction in which coastal engineering is now moving. Indeed there has been a steady increase in interest in coastal management generally, particularly in environmental and planning circles.

The subject has also occupied the attention of the Government and this Conference has followed the publication of the draft Planning Policy Guidance note (PPG) on Coastal Planning by the Department of the Environment and also the report on Coastal Zone Protection and Planning by the Environment Committee of the House of Commons. There has also been a press release on the National Audit Office investigation into costs of coastal defence.

Tim Cox, the Planning Officer for Sefton Metropolitan Borough, and Lynette Leeson, SE Regional Officer of the Countryside Commission, kindly agreed to join the five engineers on the Conference Organizing Committee in order to assist in the selection of appropriate subjects and authors. From the list of delegates it appears that the objective of attracting a wider spread of disciplines has been achieved in some measure.

There is a universal awareness of the multiplicity of both responsibilities and interests in the coastal zone which inevitably lead to conflict. The clear message from the papers and discussion is that problems must be solved by liaison and co-ordination of effort between engineers, planners and conservationists.

CLOSING ADDRESS

Although some progress has undoubtedly been made in the direction of consultation, particularly with the coastal cell groups, there is still a serious lack of co-ordination and early consultation in many areas, especially on a regional basis. The importance of early liaison on conservation issues has been made very apparent.

Tim Cox makes the point effectively in Paper 16: 'Management experience in coastal areas has one clear message: the need for integration - a real pooling of effort which reflects at professional and political level the rich interlocking of concerns and actions. Professional interests are helpful in the skills they provide but unhelpful in the barriers that can and do emerge. Active interdisciplinary work is needed which does not seek to define an exclusive role for the ecologist, planner or engineer but allows each to appreciate the contribution of others in constructive and productive interaction.'

While the PPG note was welcomed as a first step, there has been a general consensus view expressed that it falls far short of what is needed to achieve integrated Coastal Zone Management (CZM). It has perhaps a rather narrow definition of the coastal zone, deals mainly with land-based issues and has no new mechanisms for resolving conflicts. It is, of course, limited by exhausting legislation. The conclusions of the Select Committee report are interesting and reflect some of the views expressed at this Conference. It will be interesting to follow the Government's reaction to the report.

There appears to be a widespread view that there must be a review and rationalisation of responsibilities under a new legal framework. This would have to cover current planning exemptions such as ports and inshore areas beyond low water mark. Other speakers have expressed the view that legislation will take too long and that statutory powers may not in the end succeed. They would prefer to achieve the same ends by co-operation under the existing statutory framework.

Whatever route is taken, the ultimate aim is a holistic approach to CZM and the establishment of both national and regional strategies. These must take a balanced view of all interests on a short-term and long-term basis. The concept of CZM structure plans possibly at both regional and local level has received strong support and is an essential management tool.

In the event of rationalisation of the legal framework the opportunity might be taken to review planning controls in the light of the American and Australian experience. Statutory CZM has been practised in Australia for many years.

While engineers have in the past found themselves in a design straitjacket created by historic coastal structures and statutory frameworks, there is a progressive move to softer options which are more sympathetic to the coastal environment. There is now a proper appreciation of the need for coastal data collection and monitoring on a comprehensive scale as an essential design and management tool. It appears that opportunities for funding such work may be improving.

Sophisticated coastal management procedures and tools are now being developed, including GIS, and have been well described. There is, however, still an overall lack of understanding of many of the complex coastal processes, particularly nearshore and offshore sediment movement, and this must remain an important area for future research. This is linked to the need for accurate sediment budgets to be established without which reliable management decisions cannot be made. These may include the establishment of sacrificial frontages.

The topic of sediment budgets leads on to the clear future need for large quantities of beach nourishment material, especially with the prospect of rising sea levels. The competition between such needs and the demand for concrete aggregates requires the consideration of alternative quality materials and locations.

The challenge and uncertainties of potential sea level rise and increased storminess have not been specifically addressed at this Conference. However, the symposium organised by the Institution of Civil Engineers in October 1992 on Engineering in the Uncertainty of Climatic Change includes three papers on coastal engineering issues as well as papers on climate change scenarios by speakers from the Met Office and the NERC.

The issues of sea level rise and changes in storminess are, of course, vital management ingredients involving the need for flexibility and choice of options until such time as changes become more predictable. Radical solutions, including set-back and abandonment, will have to be considered as part of any overall regional strategy.

Estuaries, in particular, have been identified as a matter of great environmental importance and concern at this Conference. The progressive movement of port installations to deeper water, pollution and the gradual encroachment from land-claim have taken their toll. The prospect of sea level rise constitutes a further threat.

The need for a lead authority in CZM has been emphasised but it remains unclear who this could be. The complexity of the problem is acknowledged and it is recognised that no single body could be charged with the execution of all the relevant functions.

Perhaps the lead authority should give the framework and guidance within which existing bodies could function in a co-operative manner. The NRA was suggested as a possible lead authority but the lack of public accountability was seen as a problem.

The coastal groups have also been cited as a possible mechanism to drive CZM but some have felt that they might be too fragile and unstable. Furthermore, different groups are set up in different ways, so that to fulfil a national role they would have to be constituted formally on a common basis.